Palgrave Studies in Green Criminology

Criminologists have increasingly become involved and interested in environmental issues to the extent that the term Green Criminology is now recognised as a distinct subgenre of criminology. Within this unique area of scholarly activity, researchers consider not just harms to the environment, but also the links between green crimes and other forms of crime, including organised crime's movement into the illegal trade in wildlife or the links between domestic animal abuse and spousal abuse and more serious forms of offending such as serial killing. This series will provide a forum for new works and new ideas in green criminology for both academics and practitioners working in the field, with two primary aims: to provide contemporary theoretical and practice-based analysis of green criminology and environmental issues relating to the development of and enforcement of environmental laws, environmental criminality, policy relating to environmental harms and harms committed against non-human animals and situating environmental harms within the context of wider social harms; and to explore and debate new contemporary issues in green criminology including ecological, environmental and species justice concerns and the better integration of a green criminological approach within mainstream criminal justice. The series will reflect the range and depth of high-quality research and scholarship in this burgeoning area, combining contributions from established scholars wishing to explore new topics and recent entrants who are breaking new ground.

Joshua Ozymy · Melissa Jarrell Ozymy

The U.S. Administrative State and the Protection of Environmental Crime Victims

palgrave
macmillan

Joshua Ozymy
Department of Criminal Justice
University of Nevada, Reno
Reno, NV, USA

Melissa Jarrell Ozymy
Department of Criminal Justice
University of Nevada, Reno
Reno, NV, USA

ISSN 2946-269X ISSN 2946-2703 (electronic)
Palgrave Studies in Green Criminology
ISBN 978-3-031-62679-1 ISBN 978-3-031-62680-7 (eBook)
https://doi.org/10.1007/978-3-031-62680-7

Cover illustration: Cavan Images/Getty Images

This Palgrave Macmillan imprint is published by the registered company Springer Nature Switzerland AG
The registered company address is: Gewerbestrasse 11, 6330 Cham, Switzerland

Competing Interests The authors have no competing financial and/or non-financial interests in relation to the content of the book.

Ethics Approval All of the analysis undertaken in the manuscript involves exploration of publicly available, secondary data and presents no issues of informed consent among human participants or ethical standards disclosures for proper treatment of non-human animals in scientific research.

Preface

The U.S. EPA typically ensures compliance with the law through a host of administrative and civil tools. Despite these efforts, companies and individuals persist in committing serious violations of law that victimize the public, causing an unknown amount of economic damage, injuries, and even death. Such transgressions require EPA to engage in criminal investigations and collaboration with the U.S. Department of Justice to prosecute offenders. Yet we still know very little about the universe of victims of environmental harm, how offenders are sanctioned, and whether and how victims are made whole in the United States. Relying on a novel analysis of EPA criminal investigations that lead to prosecution from 1983 to 2022, this book provides the first comprehensive analysis of environmental crime victims within criminal prosecutions in the United States. Combining empirical analysis with narrative discussion of numerous criminal prosecutions, it provides novel insights to help advance a stronger empirical understanding of victims of environmental harm, the punishment of environmental offenders, and aids in furthering the development of an environmental victimology.

Reno, USA

Joshua Ozymy
Melissa Jarrell Ozymy

Contents

1	**Introduction**	1
	Environmental Crime Victims as the Nexus Between Policy and Criminology	3
	Methodological Approach and Limitations	4
	Conclusion	7
	Notes and References	8
2	**Cops on the Beat and Victims in the Street**	13
	Deterrence Theory, Compliance, and Enforcement	14
	Environmental Law Enforcement in Europe	18
	Environmental Victimology	19
	The Evolution of Criminal Enforcement	22
	Conclusion	26
	Notes and References	27
3	**Employees as Victims**	43
	Patterns in Prosecutions	44
	Themes in Prosecutions	49
	Hazardous Substances Crimes	49
	Air Pollution Crimes	51
	Water Pollution Crimes	52
	State-Level Environmental Crimes	53
	Punishment and Significant Prosecutions	54
	Conclusion	60
	Notes and References	61

4 When Companies Victimize the Public 69
Patterns in Prosecutions 70
Themes in Prosecutions 74
Hazardous Substances Crimes 75
Water Pollution Crimes 77
Air Pollution Crimes 78
State-Level Environmental Crimes 79
Punishment and Significant Prosecutions 79
Conclusion 83
Notes and References 84

5 When Individuals Victimize the Public 89
Patterns in Prosecutions 90
Themes in Prosecutions 93
Hazardous Substances Crimes 94
Water Pollution Crimes 97
Punishment and Significant Prosecutions 98
Conclusion 102
Notes and References 103

6 Toward a Victimology of Environmental Crime Victims 105
Typologizing Victims of Environmental Crime 106
How They Were Victimized 108
Punishing Environmental Offenders 112
Possible Victims 115
Environmental Crimes Resulting in Economic Harm 120
Conclusion 121
Notes and References 122

7 Protecting Victims of Environmental Crime 125
Key Findings 126
Trends in Prosecutions and Prospects for Deterrence 127
Obstacles to Stronger Criminal Enforcement 129
Prospects for Change 131
Notes and References 134

Index 141

About the Authors

Joshua Ozymy is a Professor of Criminal Justice in the School of Social Research and Justice Studies at the University of Nevada, Reno. He was previously a Professor of Political Science at the University of Tennessee at Chattanooga and Texas A&M University-Corpus Christi, where he was also Director of the Honors Program and Director of Strategic Initiatives. His current research centers on how and why we enforce environmental laws in the United States, with a particular focus on criminal enforcement.

Melissa Jarrell Ozymy is a Professor and Chair of the Department of Criminal Justice in the School of Social Research and Justice Studies at the University of Nevada, Reno. She was previously Department Head of Social, Cultural and Justice Studies at the University of Tennessee at Chattanooga and both Dean of University College and Chair of the Department of Undergraduate Studies at Texas A&M University-Corpus Christi. She studies environmental crime, victimization, and justice.

List of Figures

Fig. 3.1	Number of prosecutions with employees as environmental crime victims by EPA fiscal year, 1983–2022 (*Source* EPA Summary of Criminal Prosecutions Database)	45
Fig. 3.2	Number of prosecutions with employees as environmental crime victims by U.S. State, 1983–2022 (*Source* EPA Summary of Criminal Prosecutions Database)	45
Fig. 3.3	Charging patterns in prosecutions with employees as environmental crime victims, 1983–2022 (*Source* EPA Summary of Criminal Prosecutions Database)	46
Fig. 3.4	Patterns in criminal behavior for prosecutions with employees as environmental crime victims, 1983–2022 (*Source* EPA Summary of Criminal Prosecutions Database)	47
Fig. 3.5	Categories of victimization for employees that were victims of environmental crime (*Source* EPA Summary of Criminal Prosecutions Database)	47
Fig. 3.6	Themes in prosecutions with employees as environmental crime victims (*Source* EPA Summary of Criminal Prosecutions Database)	50
Fig. 4.1	Number of prosecutions where companies victimized the public by EPA fiscal year, 1983–2022 (*Source* EPA Summary of Criminal Prosecutions Database)	70
Fig. 4.2	Number of prosecutions where companies victimized the public by U.S. State and Territory, 1983–2022 (*Source* EPA Summary of Criminal Prosecutions Database)	71

Fig. 4.3 Charging patterns in prosecutions where companies victimized the public, 1983–2022 (*Source* EPA Summary of Criminal Prosecutions Database) 72
Fig. 4.4 Patterns in criminal behavior in prosecutions where companies victimized the public, 1983–2022 (*Source* EPA Summary of Criminal Prosecutions Database) 72
Fig. 4.5 Categories of victimization in prosecutions where companies victimized the public (*Source* EPA Summary of Criminal Prosecutions Database) 73
Fig. 4.6 Themes in prosecutions where companies victimized the public (*Source* EPA Summary of Criminal Prosecutions Database) 74
Fig. 5.1 Number of Prosecutions where Individuals Victimized the Public by EPA Fiscal Year, 1983–2022 (*Source* EPA Summary of Criminal Prosecutions Database) 90
Fig. 5.2 Number of prosecutions where individuals victimized the public by U.S. State and Territory, 1983–2022 (*Source* EPA Summary of Criminal Prosecutions Database) 91
Fig. 5.3 Charging patterns in prosecutions where individuals victimized the public, 1983–2022 (*Source* EPA Summary of Criminal Prosecutions Database) 91
Fig. 5.4 Patterns in criminal behavior in prosecutions where individuals victimized the public, 1983–2022 (*Source* EPA Summary of Criminal Prosecutions Database) 92
Fig. 5.5 Categories of victimization in prosecutions where individuals victimized the public (*Numbers are rounded. *Source* EPA Summary of Criminal Prosecutions Database) 93
Fig. 5.6 Themes in prosecutions where individuals victimized the public (*Source* EPA Summary of Criminal Prosecutions Database) 94
Fig. 6.1 Typology of environmental crime victims (*Source* EPA Summary of Criminal Prosecutions Database) 106
Fig. 6.2 Categories of environmental crimes (*Source* EPA Summary of Criminal Prosecutions Database) 108
Fig. 6.3 Charging patterns in prosecutions with environmental crime victims (*Source* EPA Summary of Criminal Prosecutions Database) 110
Fig. 6.4 Most common offender behaviors (*Source* EPA Summary of Criminal Prosecutions Database) 110
Fig. 6.5 The consequences of victimization (*Source* EPA Summary of Criminal Prosecutions Database) 112

Fig. 6.6 Common offenses against potential environmental crime victims (*Source* EPA Summary of Criminal Prosecutions Database) 116
Fig. 7.1 Number of prosecutions with environmental crime victims by EPA Fiscal Year, 1983–2022 (*Source* EPA Summary of Criminal Prosecutions Database) 128
Fig. 7.2 Number of prosecutions with environmental crime victims by U.S. State and Territory, 1983–2022 (*Source* EPA Summary of Criminal Prosecutions Database) 129

List of Tables

Table 3.1	Total penalties assessed at sentencing for prosecutions with employees as environmental crime victims	55
Table 3.2	Cases of significant probation assessed at sentencing for prosecutions with employees as environmental crime victims	55
Table 3.3	Cases of significant incarceration assessed at sentencing for prosecutions with employees as environmental crime victims	57
Table 3.4	Cases of significant monetary penalties assessed at sentencing for prosecutions with employees as environmental crime victims	58
Table 3.5	Percentage of cases where restitution was awarded to victims for prosecutions with employees as environmental crime victims	59
Table 4.1	Total penalties assessed at sentencing for prosecutions where companies victimized the public	80
Table 4.2	Cases of significant probation assessed at sentencing for prosecutions where companies victimized the public	81
Table 4.3	Cases of significant incarceration assessed at sentencing for prosecutions where companies victimized the public	81
Table 4.4	Cases of significant monetary penalties assessed at sentencing for prosecutions where companies victimize the public	82

Table 4.5 Percentage of cases where restitution was awarded to victims for prosecutions with where companies victimized the public 83
Table 5.1 Total penalties assessed at sentencing for prosecutions where individuals victimized the public 99
Table 5.2 Cases of significant probation assessed at sentencing for prosecutions where individuals victimized the public 99
Table 5.3 Cases of significant incarceration assessed at sentencing for prosecutions where individuals victimized the public 100
Table 5.4 Cases of significant monetary penalties assessed at sentencing for prosecutions where individuals victimized the public 101
Table 5.5 Percentage of cases where restitution was awarded to victims for prosecutions with where individuals victimized the public 102
Table 6.1 Total penalties assessed at sentencing for prosecutions with environmental crime victims 112
Table 6.2 Percentage of cases where restitution was awarded to environmental crime victims 114

CHAPTER 1

Introduction

Abstract While there is a small, but growing literature that tries to tackle the thorny question of whether criminal enforcement tools for the environment are effective in the United States, few look at the role of environmental crime victims within the enforcement process. For political science and public policy researchers that care about the consequences of enforcement to criminologists that want to develop empirical and theoretical understandings of victimization, to the general public and policymakers concerned with how the administrative state protects the public from environmental harm, such an investigation is warranted. While earlier works addressed a number of issues about these questions, they have yet to fully explore the role environmental crime victims play in criminal investigations and prosecutions, the punishments meted out to offenders that harmed them, and to fully define and typologize the universe of these victims in the United States.

Keywords Environmental crime victims · Scope and method · Limitations

J. Ozymy and M. Jarrell Ozymy, *The U.S. Administrative State and the Protection of Environmental Crime Victims*, Palgrave Studies in Green Criminology, https://doi.org/10.1007/978-3-031-62680-7_1

On February 3, 2023, a freight train operated by the Norfolk Southern Railway derailed outside of East Palestine, Ohio.[1] Of the 50 cars that tumbled from the tracks or caught fire, it appears that 10 carried hazardous materials, including five that contained vinyl chloride, a highly volatile gas that ignited causing an incendiary toxic plume that threatened the residents nearby.[2] About 4,700 residents were encouraged to evacuate the area and an evacuation order was issued within a one-mile radius.[3] The U.S. Environmental Protection Agency (EPA) conducted air monitoring of the area and the Ohio Department of Health and the U.S. Department of Health and Human Services opened a Health Assessment Clinic in the area, as did Norfolk Southern, to address the immediate public health implications of this environmental disaster.[4]

On February 22, 2023, the EPA ordered Norfolk Southern to pay for all cleanup associated with the derailment, beginning the first phase in a longer-term remediation process, while offering free indoor air monitoring and cleaning services to families and businesses.[5] While air monitoring and house cleaning were nice gestures, public health experts and members of the local community were greatly worried about the short-term and long-term health impacts of the exposure to the toxic fumes and fallout of the vinyl chloride plume. Residents were right to be worried. Long term exposure to vinyl chloride has been shown to cause a variety of cancers and other health problems.[6]

In the coming months, the EPA, along with a number of other agencies, such as the National Transportation Safety Board (NTSB), the Ohio EPA, and likely the Occupational Safety and Health Administration (OSHA), will conduct investigations to find the cause of the accident and the attention given to it by the media will wane significantly.[7] The EPA issued a General Notice of Liability from the EPA Region 5 office in Ohio, informing the company that the agency, under their authority provided by the Comprehensive Environmental Response, Compensation, and Liability Act (CERCLA) believes the company may be responsible for reimbursing EPA's cleanup costs for the site.[8] EPA Director, Michael S. Regan, noted publicly that the EPA could fine the company up to $70,000 dollars a day.[9]

Environmental Crime Victims as the Nexus Between Policy and Criminology

The EPA may choose to take no further punitive actions toward the company. If the company is found to have violated federal environmental law, the typical approach taken by EPA is to bring the transgressor back into compliance with the law through administrative or civil enforcement tools.[10] However, when violations involve culpable conduct or result in significant harm, the EPA can pursue a criminal investigation and collaborate with prosecutors within the U.S. Department of Justice (DOJ) to pursue criminal prosecution of a potential environmental offender.[11]

After the EPA's founding in 1970, it was not long before regulators realized many individuals and companies will choose to willfully break environmental laws, absent mechanisms to punish and deter such crimes.[12] Adding criminal enforcement tools to the environmental enforcement structure ramped up in earnest by the 1980s; resources to police environmental crimes were institutionalized at EPA, funding to prosecute environmental crimes increased at the DOJ, and Congress upgraded criminal penalties for many environmental violations.[13] By giving resources to EPA and DOJ to investigate and prosecute environmental crimes and enhancing criminal provisions in environmental law to carry stiff punishments, up to incarceration, Congress acknowledged the growing problem of environmental crime and the need for stronger deterrent mechanisms to manage it.[14] EPA and DOJ have used their criminal enforcement authority in a number of past situations that are inherently similar in many respects to the Norfolk Southern case. Some examples that are useful for context here include the prosecution of Trinity Railcar Repair, Techsol Company, and Explo Systems.[15]

Trinity Railcar Repair caused an ammonia leak from a railcar located in Fort Edward, New York, causing some 400 people to be evacuated. The company was charged with violations of the U.S. Clean Air Act (CAA) and was sentenced to pay a $125 assessment, $200,000 fine, and $41,372 in restitution, as well as make $200,000 in payments.[16] On October 28, 2004, 22,000 gallons of coal tar spilled at the Techsol Chemical Company that leaked from a railroad tanker into the city sewer system and 500 people were evacuated from their homes. The company's president, James Richard Holt, pleaded guilty to violations of the U.S. Resource Conservation and Recovery Act (RCRA) and violations of the U.S. Clean Water Act (CWA) for storage or disposal of hazardous waste without a permit and illegal discharge to the navigable waters of the United States. Holt was sentenced to serve 6 months of home detention, 60 days community confinement, 60 months of probation, and to pay $13,948 in restitution.[17] On October 15, 2012, an explosion occurred at Camp Minden, causing an evacuation of the town of Doyline, Louisiana, shattering windows, destroying a bunker, and derailing 11 rail cars. Officials from Explo Systems conspired to defraud the federal government by gaining lucrative contracts from the U.S. Army to demilitarize propellants, but instead sent them for illegal disposal or illegally stored them, which led to the explosion. Co-owner David Alan Smith was sentenced to 55 months in prison, three years of supervised release, and to pay $34,798,761 in restitution.[18] Charles Ferris Callihan, Director of Support Technology, was sentenced to 24 months in prison, one year of supervised release, and $207,599 in restitution.[19]

While significant violations may or may not have taken place in the East Palestine Derailment that warrant criminal enforcement actions from EPA, the local community was victimized as the result of Norfolk Southern's actions. The immediate victimization included burning eyes and rashes, relocation, and closing of schools and other facilities. The longer-term risks from toxic exposure are unknown, but could include injuries, respiratory illness, or even death.[20]

When we study the effects of environmental policy implementation, we want to explore questions about how government agencies, such as EPA and DOJ, enforce the law, as well as the effects of those actions. If we wish to examine the broader consequences of environmental violations to human populations, particularly those that are victims of environmental crime, we might employ a green criminological lens to frame our investigation. Exploring the role of environmental crime victims within the enforcement process in the United States opens up an incredibly fascinating set of questions, prescient and germane to both literatures that have mostly been unexplored in a systematic manner over time.

Methodological Approach and Limitations

Our primary data source comes from the EPA's Summary of Criminal Prosecutions Database.[21] The database contains all EPA environmental crime investigations that lead to prosecution since 1983. The database contains a rich data source in the form of textual summaries of each prosecution. We searched the database by EPA fiscal year from the first case in the data that was adjudicated in 1983 through the end of fiscal year 2022. In total, we cataloged 2,828 prosecutions for our analysis. We examined each of these prosecutions for examples where the text summaries mention human victims of the environmental crime in question and identified 117 prosecutions for the analysis herein.

In order to extract consistent meaning from the textual case summaries in the EPA database, we used content analysis as our analytical technique. This approach allowed us to explore the case summaries with an eye toward organizing the data to determine the relationship between certain words and phrases to extrapolate broader themes and derive a meaningful understanding of the data.[22] Ours is not an example of a complex content analysis that attempts to derive deep meaning from disparate data, but to systematically decide which aspects of the case summaries are meaningful for understanding the role of victims in environmental crime prosecutions. From the outset, our general goals were to explore questions such as: who were the offenders, what were the categories of victims, how they were victimized, how were offenders punished, were victims made whole and how?

With these basic questions in mind that would inform the genesis of the manuscript, we developed protocols to reliably and consistently code the data. Because of our extensive experience working with this database, we had a strong baseline understanding of what we sought to gather, and how to collect and interpret the data. We employed a previous technique from our past research, where we created coding protocols by piloting them for roughly a month with two coders gathering data from the database. Each individual gathered the data independently, but on a similar timeline. We then met to find consensus on any diverging values in each of the codebooks. Inter-coder reliability, defined as the number of agreed upon items divided by the number of non-agreed items, was approximately 95% for our dataset.[23]

We built on this approach, by working through all of the prosecutions to find cases with identifiable victims. Once we agreed upon cases that met this basic criteria, we captured data on a number of relevant characteristics for each prosecution summary. Using an Excel spreadsheet, we gathered the following data from each case: name of primary defendant, number of defendants, state identifier (for the prosecution), docket number, short narrative of the case, whether a company was a named defendant in the prosecution, the presence of additional criminal charges, such as obstruction, false statements and fraud in the case, charging patterns including violations of major environmental laws, whether a defendant was charged with state-level violations, and penalties assessed at sentencing. We aggregated penalties across companies and individual defendants in each case, including total months of probation, total months of incarceration, total hours of community service, and monetary penalties including fines, restitution, special payments or assessments, and any other monetary penalties noted in the prosecution summaries. We also collected data from the summaries on restitution amounts provided to victims and their families. Our analysis of 2,828 prosecutions leads us to center our analysis for this manuscript on 117 prosecutions we agreed involved identifiable human victims of environmental crime. To explore their role in environmental crime prosecutions, we analyzed the data from our content analysis that allowed us to demonstrate the universe of these victims, including where and when they were victimized, how they were victimized, and describe punishments and restitution.

There are a few limitations in our analytical approach. First, we chose to not include any additional sources of data outside of the information listed in the EPA's database. We felt that using other sources of data that were not completely consistent across prosecutions would likely jeopardize our ability to generate a reliable and valid source of data for analysis. If a prosecution was not listed in the EPA's database, then it would be unknown to us and absent from our analysis. Likewise, because we use the database exclusively, it may not represent the totality of every environmental crime prosecution in the United States over time and our ability to generalize from our findings is limited by this approach. Third, we can only speak to victimization in the context of prosecutions and not victimization more generally that occurs outside that context. Fourth, our data and analysis are limited to the United States, restricting our ability to generalize or speak to issues of victimization in other countries or from a cross-national approach. Finally, our analysis has the temporal limitation of ending in EPA fiscal year 2022 and the analysis could become dated with time.

While all of the above represent limitations on how we can speak about victims of environmental crime both in the United States and abroad, they do not represent structural deficiencies in our approach or hinder our ability to illuminate important questions about such victims. There is arguably no better and more consistent dataset on the prosecution of environmental crime in the United States than the EPA's official database. There is also no centralized database nor current method of constructing a consistent database of all the victims of environmental crime over space and time in the United States and certainly not a comparable dataset with indicators that can be compared across countries. Contextualizing these complex cases of victimization would be difficult across a broad spectrum of place and time as would constructing a valid database, where one could engage in a similar level of analysis as we have herein. Furthermore, our approach allows us to move past the thornier question of whether and how people are victimized by corporations and other individuals, by focusing on legal prosecutions where we have identifiable victims, offenders, and outcomes for both, which is a significant asset in our approach. Our analysis is also easily buildable for future studies that wish to further analyze environmental crime victims using the database years, if not decades in the future. As long as EPA maintains the database and given all agencies like to highlight their accomplishments and they have maintained it now for some four decades, it seems more likely than not this data source will be updated for many years.

The strongest contribution of our analysis, despite its limitations, comes not only from the basic description of crimes and victims, but how our content analysis of the prosecutions allowed for a deeper dive into the data. We were able to answer structural questions about environmental victimization in the United States, by exploring broader themes and positing answers to a number of important questions such as: What were the most common offenses that lead to victimization? What types of environmental crimes resulted in victims? What were the broader categories of victimization? What were the outcomes in terms of how people were victimized? The lack of data and specialized nature of studying victims of environmental crime have left many of these questions systematically unanswered in the United States and abroad. While we cannot speak to the latter and are unable to answer all questions about the former, our method and approach allow us to utilize a valid, consistent, and robust source of data to proffer, in our opinion, strong and data-driven answers to these important questions left open for discussion and analysis in the literature.

Conclusion

While scholars and journalists may make claims regarding the type and extent of environmental crime victims within the United States and abroad, it is difficult to know the extent of victimization, the type, or how offenders are punished—a victimology of environmental crime victims is warranted. This book is an effort to explore the universe of environmental crime victims through the novel approach of exploring their role in environmental crime prosecutions. Public policy researchers that study environmental policy implementation can also benefit from a more significant treatment of the effects of implementing policies. Studying these effects often encompasses broader trends in pollution reduction, the adoption and success of technical administrative solutions to problems with pollution, and the universe of civil enforcement efforts.[24] Very few scholars have examined the important nexus between criminal enforcement at EPA and the prosecution of offenders by DOJ, a relationship that requires significant collaboration and crosses disciplinary boundaries. Exploring the role of victims in environmental crime prosecutions, organizing this universe, and working to generalize our findings into a broader framework for understanding this phenomenon benefit the study of victimology, as well as public policy. We hope to provide a substantive treatment of the subject herein.

We move forward in Chapter 2, with a literature review that describes the history and evolution of the criminal enforcement process for the environment in the United States, followed by a discussion of the literature on environmental crime victims, green criminology, and criminal deterrence. We are interested in exploring how criminal enforcement evolved, our empirical and theoretical knowledge about environmental crime victims, and how offenders are punished for their transgressions. Upon analyzing our data, we found that victims could be placed within the categories of individuals victimized at work, individuals victimized by companies, and individuals victimized by others. We organized our analysis in Chapters 3, 4, and 5, respectively, by these categories. In Chapter 6, we bring the previous analysis together and describe broad trends in victimization and offending to develop a victimology of crime victims within criminal prosecutions. In Chapter 7, we conclude with a summary of our key findings, followed by a discussion of the obstacles to stronger protections for victims within the administrative state, and then recommendations for institutional changes to buffer these obstacles.

Notes and References

1. Morgan Winsor and Meredith Desilo, 2023, Ohio Train Derailment: NTSB Chair Issues Plea to Those Spreading Misinformation,

ABC News, Available from: https://abcnews.go.com/US/ohio-train-derailment-ntsb-chair-issues-plea-spreading/story?id=97281671.

2. Li Cohen, 2023, What Is Vinyl Chloride? Here Are Some of the Hazardous Materials on the Train that Derailed in Ohio, CBS News, Available from: https://www.cbsnews.com/philadelphia/news/what-is-vinyl-chloride-what-other-chemicals-were-on-train-derailed-east-palestine-ohio-paulsboro-new-jersey/.
3. Peter Charalambous and Morgan Winsor, 2023, Ohio Train Derailment: Evacuation Order Lifted, Residents can Return Home, ABC News, Available from: https://abcnews.go.com/US/ohio-train-derailment-residents-forced-evacuate-allowed-return/story?id=96941403.
4. Julia Jacobo, 2023, Why the Toxins from the Ohio Train Derailment Could have Posed Deadly Threats for the Residents Nearby, ABC News, Available from: https://abcnews.go.com/Health/toxins-ohio-train-derailment-posed-deadly-threats-residents/story?id=96978394.

   U.S. EPA, 2024, Air Sampling Data: East Palestine, Ohio Train Derailment, Available from: https://www.epa.gov/oh/air-sampling-data-east-palestine-ohio-train-derailment.
5. U.S. EPA, 2024, East Palestine, Ohio Train Derailment Emergency Response, Available from: https://www.epa.gov/oh/east-palestine-ohio-train-derailment-emergency-response.
6. Matthew S. Fralish and John W. Downs, 2022, Vinyl Chloride Toxicity, Available from: https://www.ncbi.nlm.nih.gov/books/NBK544334/.
7. Occupational Safety and Health Administration (OSHA), 2023, About OSHA, Available from: https://www.osha.gov/aboutosha.

   Ohio Environmental Protection Agency, 2023, East Palestine Train Derailment, Available from: https://epa.ohio.gov/home.

   National Transportation Safety Board, 2023, NTSB Issues Investigative Update on Ohio Trail Derailment, Available from: https://www.ntsb.gov/news/press-releases/Pages/NR20230214.aspx.
8. U.S. EPA, 2024, Norfolk Southern Railway Company, General Notice of Liability 2-10-203, Available from: https://www.epa.gov/system/files/documents/2023-02/Norfolk%20Southern%20East%20Palestine%20Train%20Derailment%20General%20Notice%20Letter%202.10.2023%20%281%29.pdf.

Comprehensive Environmental Response, Compensation, and Liability Act, 42 USC 103 (1980). The Act authorizes EPA to find responsible parties to pay for the cost of cleaning up and remediating sites contaminated with hazardous substances.

EPA organizes its offices to oversee policy implementation, enforcement, and other issues across 10 regional offices. Ohio is in EPA Region 5 that oversees Illinois, Indiana, Michigan, Minnesota, Ohio, Wisconsin, and 35 Indian Tribal governments.

U.S. EPA, 2024, EPA Region 5, Available from: https://www.epa.gov/aboutepa/epa-region-5.

U.S. EPA, 2024, Region 5 Tribal Program, Available from: https://www.epa.gov/tribal/region-5-tribal-program.

9. Nouran Salahieh, Holly Yan, and Claire Colbert, 2023, EPA Says It Can Fine Norfolk Southern $70,000 a Day If It Falls Short of Cleaning Up and Paying for the Ohio Toxic Train Wreck, CNN, Available from: https://www.cnn.com/2023/02/22/us/ohio-train-derailment-east-palestine-wednesday.
10. U.S. EPA, 2024, Basic Information on Enforcement, Available from: https://www.epa.gov/enforcement/basic-information-enforcement.
11. Earl E. Devaney, 1994, The Exercise of Investigative Discretion, 3–4, Available from: https://www.epa.gov/sites/production/files/documents/exercise.pdf.

    David M. Uhlmann, 2009, Environmental Crime Comes of Age: The Evolution of Criminal Enforcement in the Environmental Regulatory Scheme, 4 Utah Law Review, 1224.

    David M. Uhlmann, 2014, Prosecutorial Discretion and Environmental Crime, 38 Harvard Environmental Law Review, 159.

    U.S. EPA, 2024, Criminal Enforcement Overview, Available from: https://www.epa.gov/enforcement/criminal-enforcement-overview.
12. Robert I. McMurry and Stephen D. Ramsey, 1986, Environmental Crime: The Use of Criminal Sanctions in Enforcing Environmental Laws, 19 Loyola of Los Angeles Law Review, 1136–1141.
13. Richard J. Lazarus, 1994, Assimilating Environmental Protection into Legal Rules and the Problem with Environmental Crime, 27 Loyola Law Review, 867–870.

14. Robert I. McMurry and Stephen D. Ramsey, 1986, Environmental Crime: The Use of Criminal Sanctions in Enforcing Environmental Laws, 19 Loyola of Los Angeles Law Review, 1136–1141.
15. Trinity Railcar Repair, Inc. (N.D. New York 1:02-CR-444-001, 2003); James Richard Holt (S.D. West Virginia 3 09 CR 00231, 2011); Charles Ferris Callihan (W.D. Louisiana 16-CR-00214-06, 2019).
16. Legal cases that are discussed throughout the book, particularly the analysis sections, are those cases involving victims that were accessed through the U.S. EPA's *Summary of Criminal Prosecutions Database*. We use this database exclusively to gather all of our data for the analysis herein. Legal citation is taken from the database and resembles the Trinity Railcar Repair, Inc. example here:

    Trinity Railcar Repair, Inc. (N.D. New York 1:02-CR-444-001, 2003).

    U.S. EPA, 2024, Summary of Criminal Prosecutions Database, Available from: https://www.epa.gov/enforcement/summary-criminal-prosecutions.

    U.S. Clean Air Act (CAA), Pub. L. No. 91-604 (1970). The CAA Extension of 1970 (P.L. 91-604) gives EPA authority to regulate emissions from mobile and point-source emissions.
17. James Richard Holt (S.D. West Virginia 3 09 CR 00231, 2011).

    Clean Water Act (CWA), 33 U.S.C. §1251 et seq. (1972), the Federal Water Pollution Control Act Amendments of 1972 (P.L. 92-500), gives EPA authority to create a system of managing discharges from the navigable waters of the United States.

    U.S. Resource Conservation and Recovery Act (RCRA): 42 U.S.C. §6901 et seq. (1976), gives EPA authority to regulate solid waste and hazardous waste from cradle to grave.
18. United States Attorney's Office, 2018, Explo Officials Sentenced for Roles in Conspiracy That Led to Illegal Dumping of Munitions and Explosion at Camp Minden, Available from: https://www.justice.gov/usao-wdla/pr/explo-officials-sentenced-roles-conspiracy-led-illegal-dumping-munitions-and-explosion.
19. Charles Ferris Callihan (W.D. Louisiana 16-CR-00214-06, 2019).
20. Scripps News, 2023, One of the Few Researchers Studying Vinyl Chloride's Health Effects, Available from: https://www.msn.com/en-us/health/medical/one-of-the-few-researchers-studying-vinyl-chlorides-health-effects/vi-AA17CWrO.

21. U.S. EPA, 2024, Summary of Criminal Prosecutions Database, Available from: https://www.epa.gov/enforcement/summary-criminal-prosecutions.
22. Ole R. Holsti, 1969, Content Analysis for the Social Sciences and Humanities, 140. Addison Wesley.
23. C. O'Connor and H. Joffe, 2020, Intercoder Reliability in Qualitative Research: Debates and Practical Guidelines, International Journal of Qualitative Methods, 19.
24. Joshua Ozymy, Bryan Menard, and Melissa L. Jarrell, 2021, Persistence or Partisanship: Exploring the Relationship Between Presidential Administrations and Criminal Enforcement by the U.S. Environmental Protection Agency, 1983–2019. 81 (1) Public Administration Review, 49–63.

CHAPTER 2

Cops on the Beat and Victims in the Street

Abstract Our goal in this chapter is to explore three aspects of environmental victimization. In the first, we provide an overview of the literature on theories of deterrence and compliance in regulatory enforcement in the United States and Europe. Second, we describe the literature on environmental/green victimology, looking broadly at victimization from a green criminological perspective that focuses on victimization through a social harm lens. Finally, we describe the historical evolution of criminal enforcement tools for the environment in the United States. We aim to provide a broad overview of the literatures in these areas to contextualize how we theorize about methods of environmental crime control, victimologies of environmental crime, and the evolution of criminal enforcement tools for the environment in the United States.

Keywords Deterrence theory · Victimology · Criminal enforcement

Theoretical frameworks for understanding how to prevent victimization stemming from serious environmental crimes focus on the calculus of decision making for potential offenders. Why do certain individuals and companies decide to violate environmental statutes and victimize other people? What regulatory responses are most appropriate to prevent

J. Ozymy and M. Jarrell Ozymy, *The U.S. Administrative State and the Protection of Environmental Crime Victims*, Palgrave Studies in Green Criminology, https://doi.org/10.1007/978-3-031-62680-7_2

victimization? How can the regulatory response be such that it deters both individuals and organizations from committing environmental crimes, on a specific basis and on a broader societal or industry level?

Deterrence Theory, Compliance, and Enforcement

The primary regulatory framework in the United States for understanding and managing violations of environmental law is compliance, where EPA or state agencies (when appropriate) attempt to bring an individual or company back into compliance with the law and when possible, mitigate the effects of pollution caused by the violator.[1] Some scholars have argued that this approach, given it attempts to return the violator into compliance and return a community or individual to a prior state before the harmful pollution occurred, also contains a restorative justice element to many regulatory actions.[2] Criminal enforcement is another approach environmental law enforcement takes for crimes involving significant harm and culpable conduct.[3] For those violations that are very serious and/or represent negligent or knowing violations of law, EPA can pursue a criminal investigation and work with DOJ prosecutors to potentially file criminal charges against violators.[4]

The general approach to environmental crime enforcement centers on deterrence, rather than compliance. Deterrence theory suggests that the probability of being caught must be sufficiently high and the severity of punishment must be certain enough to directly deter offenders and pose indirect examples to the regulated community to create a general deterrence effect.[5] This is a command and control strategy that secures deterrence of crime via the threat of formal sanctions.[6] For deterrence be effective as a theoretical framework and regulatory approach to managing serious environmental crime in the United States, four variables would arguably need to hold: the availability of strong punishments, a reasonable structure for white-collar crime enforcement, workable *means rea* requirements for prosecutors to prove intent or negligence, and sufficient policing and prosecutorial resources directed toward patrolling and punishing environmental crimes.

The first requirement for deterrence of environmental crimes is that criminal violations of federal environmental law should include punishments, such as significant monetary penalties and incarceration for serious crimes. When one places another person in imminent danger of death or serious bodily injury, the crime of knowing endangerment, for example, can place the guilty party in a position to receive at sentencing up to 15 years of prison time per offense, $250,000 in fines, and up to $1 million in monetary penalties for companies.[7] Knowing endangerment holds the maximum penalty for individuals and organizations under federal environmental law, but other penalties can also be quite stiff. The movement to enhance deterrence for environmental crime began in the 1980s in the United States, when Congress upgraded penalties and began inserting felony provisions in environmental statues to enhance the deterrent effect of criminal violations of environmental law.[8] Congress was also signaling to other federal law enforcement agencies that they needed to take environmental crime seriously.[9] The potential penalties for a number of crimes are very high in the United States meeting the threshold of punishments being potentially severe enough to deter crime.

Second, for criminal enforcement to be an effective deterrent, it must be able to target white-collar criminals, particularly owners and corporate officers and managers.[10] In the United States, criminal statutes are often written to target malfeasance within an organization, allowing the creation of a deterrence mechanism for a range of white-collar criminals operating within organizations.[11] This precedent gives criminal law an added deterrence aspect that is lacking in civil and administrative law, which is arguably less effective at sanctioning white-collar criminals acting within organizations.

Third, environmental crimes, unlike street crimes, are often very hard to investigate and punish, given the difficulty of finding evidence and the cause and effect nature of environmental crimes.[12] This is hard even if the crime is acute (i.e., negligence causes a chemical explosion) and more so if it is chronic and indirect (i.e., lung disease caused by living near a petroleum refinery). Criminal law for the environment has evolved in the United States to manage this issue, by having a somewhat broader basis for establishing *mens rea* in environmental crimes. Generally speaking, establishing intent or a guilty mind is not formally necessary to prosecute criminal offenses as it would with other forms of interpersonal violence. For example, an officer or manager of an organization can be prosecuted for an environmental crime if their place within an organization makes them responsible for protecting workers or the public and their actions or lack of action resulted in pollution that injured, killed, or damaged others, in a similar manner as if they had given specific orders with intent that resulted in victimization.[13]

Finally, deterrence of environmental crimes in the United States requires a significant policing presence to patrol for potential violations and resources to prosecute crimes. On this point, the robustness of deterrence theory for environmental crimes has been challenged, due to the limited budget for environmental policing. From the 1980s on, criminal enforcement never had the resources to create a significant policing force to monitor environmental crimes across the country, nor was that ever the intent of Congress to create such a structure for the broad-based policing and prosecution of environmental crimes. Today, as it was in the mid-1990s, some 200 EPA-CID criminal enforcement staff patrol the country for environmental crimes and some 43 specialized attorneys in DOJ-ECS and the U.S. Attorney's Office prosecute crimes.[14] The lack of direct federal staff responsible for policing and prosecuting environmental crimes is certainly the linchpin in the argument that deterrence theory may not fully apply to environmental crimes.

Given its limitations in practice, due to resource restrictions to police and prosecute environmental crimes, deterrence theory may still provide a broad framework for understanding how to police and punish environmental offenders in the United States. Environmental law enforcement agencies certainly see their work through a deterrence lens and consider themselves "America's environmental crime fighters".[15] Criminal enforcement was always meant to be calculated in approach and strategic in its ambitions, targeting only the most egregious violations of law and chronic offenders.[16] Augmenting general deterrence from criminal enforcement alone is the practical reality that EPA and DOJ can pursue both criminal and civil cases against offenders to enhance any punitive or deterrent effects and can they can do so exclusively or in conjunction with state agencies.[17] Assessing the value of deterrence for explaining criminal enforcement must be done in this broader context.

To further contextualize deterrence and environmental law enforcement, it is important to reemphasize that the general principle of the U.S. regulatory approach for the environment begins with returning violators to compliance with the law, rather than subjecting them to criminal sanctions.[18] This process typically begins with administrative remedies proposed by EPA, such as the issuance of a notice of violation, providing an order of correction, or the imposition of fines.[19] These approaches can be taken by EPA or state regulators, which are often the first line of defense when an individual or organization within the regulatory community is in violation of law. If administrative remedies prove unsuccessful or the violation is serious, a variety of civil tools can be applied to help the violator regain compliance. These may include creating an environmental mitigation plan to prevent future pollution, designation of a required supplemental environmental project or a beneficial project that is part of a settlement to regain compliance, creation of an environmental monitoring plan to ensure polluting activities cease, or having a court issue a temporary or permanent injunction to stop the violator from polluting.[20] If one or more of these administrative or civil remedies proves ineffective to regaining compliance, DOJ can pursue a civil lawsuit to find a responsible party guilty and liable for damages from a polluting activity.[21]

Administrative or civil remedies can be pursued by state agencies or EPA either exclusively or in tandem as appropriate to the situation.[22] Because these agencies seek compliance as their initial regulatory response both in theory and in praxis and there is a significant amount of flexibility in how they can shape civil and administrative solutions to the situation, and because the burden of proof is lower for civil than criminal violations, and because the cost is arguably much lower than pursuing a criminal investigation and conviction, civil and administrative actions vastly outnumber criminal prosecutions historically in the United States as the primary regulatory response to environmental violations.[23]

With these factors in mind, there is still a lack of agreement on the specific or general deterrent effect of criminal prosecution for reducing environmental crimes and victimization.[24] Research shows prosecutors have sought significant penalties and pursued complex investigations historically in the United States.[25] Studies show prosecutors securing significant penalties for environmental crimes at sentencing in a variety of contexts.[26] Other studies argue that the probability of being caught and punished is so low, and that deterrence is somewhat implausible in environmental crime prosecutions.[27] Treated in isolation, our own research on the deterrent effects of criminal enforcement suggests that monitoring and punishing violations has been historically limited, but this does not rule out some specific and general deterrence effects, particularly when combined with possible civil or administrative penalties, as well as state enforcement actions.[28]

Whether criminal enforcement is effective toward the end of reducing human victimization is unknown and it is debatable whether and how it deters offenders as part of the broader administrative, civil, and criminal enforcement scheme undertaken with and alongside state agencies, the latter of which, due to data availability, have received even more limited attention in the literature.[29] The question of how enforcement may be used in cases involving human victims of environmental crime is very limited and whether it is effective for specific or general deterrence is still in dispute.[30]

Environmental Law Enforcement in Europe

Given its long history compared to other regions of developing regulatory enforcement tools to protect the environment, drawing on examples from the continent is helpful to contrast with the American approach to compliance and deterrence for environmental violations. A major starting point for a discussion focused on the regulation of the environment and managing environmental crime in Europe should begin with the main law that governs criminal sanctioning of environmental crimes, known as Directive 2008/99/EC or the Environmental Crime Directive. The Directive, adopted on October 24, 2008, gave member states until December 2010 to create criminal law penalties for environmental crime to supplement current administrative sanctioning systems.[31] The Directive requires the adoption of criminal sanctions in "72 pieces of environmental legislation" in order to deter serious violations of law.[32]

Aiming for elevating the seriousness of environmental crime as a public policy problem for both individual states and the greater European Community, the Directive set out to move environmental regulation beyond a general compliance-based system for many member states, to one that is strengthened with criminal penalties to deter environmental crimes and "demonstrate a social disapproval of a qualitatively different nature compared to administrative penalties or a compensation mechanism under civil law".[33] The Directive creates minimum rules that member states add criminal sanctions to law, acknowledge the seriousness of environmental crimes by doing so, and recognize that environmental crime often crosses borders and thus represents a serious policy problem for the greater community. This policy adoption required member states to create legal liability for serious offenses for both individuals and companies and ultimately to, "approximate criminal sanctions levels by requiring effective, proportionate and dissuasive criminal penalties for environmental crimes."[34]

Like the United States, the passage of the Directive prompted European states to graft criminal sanctions alongside existing compliance enforcement systems. The basic principles of both the European and

American regulatory systems can then be explained through regulatory models of compliance and deterrence. Differences come in how EU states enforce criminal penalties within their own unique administrative structures. Even prior to the passage of the Directive, many EU states utilized the criminal justice system to enforce environmental law, with some countries, such as UK relying more on criminal sanctions and others, such as the Netherlands, emphasizing administrative sanctions.[35] The UK, Denmark, Ireland, and Finland, as examples, often apply administrative measures to a violator first, which might then result in criminal prosecution.[36] For other member states, such as France, Belgium, Luxembourg, and the Netherlands, the appropriate agency might apply a sanction directly.[37] Germany and Austria have "quasi-criminal" systems that can apply administrative criminal sanctions and both countries, as well as Spain and Portugal, do not recognize criminal liability for corporations and must still find a way to achieve a deterrent effect with their enforcement actions.[38]

Environmental Victimology

The study of environmental victimology, also known in the literature as green victimology, is the exploration "of the social processes and institutional responses pertaining to victims of environmental crime".[39] Environmental crimes are not victimless crimes, victims simply go unnoticed, and such crimes often remain unpunished within the criminal justice system.[40] These related problems are exacerbated by governments, who often condone the actions of industry that causes green victimization and trivialize environmental crimes within the political and legal systems relative to street crimes that are seen as "real crimes" with real victims, unlike environmental crimes.[41] Consequently, our image of a victim is less likely to be that of a corporate or environmental crime and more of one that is sexually assaulted or murdered.[42] Given the nature of most environmental crimes involves chronic and often silent processes that act on victims, we rarely hear their stories or link such crimes to their victimization conceptually, in depictions of victims in the media, in surveys of the public, or the traditional study of criminology.[43] These factors bleed into a broader trivialization of environmental crimes and their victims within society, meaning the mass media rarely reports on environmental victimization unless it results in a significant and acute tragedy, such as a train derailment, refinery explosion, or other catastrophe.[44] As Matthew Hall notes, "the social, political and economic influences on the identification and recognition of 'illegal' or 'harmful' acts and omissions necessitate fundamental questions being asked of what any given criminal justice system does and does not accept as an 'environmental crime' and thus an 'environmental victim'".[45]

Understanding the limitations of the current conceptual approach that dominates the field of criminology and subsequently victimology, researchers have sought to develop a broader conceptual understanding of environmental crime from a social harm perspective.[46] Moreover, the study of environmental victimology generally focuses on the study of the victimization of humans until more recently, when scholars have increasingly turned their attention to non-human animal victims of environmental crime.[47] Using a green criminological framework, environmental or green crimes are those illegal and legal acts that may cause harm to humans, non-human animals, as well as the natural environment.[48] As a theoretical matter, green criminologists have sought to incorporate a social harm perspective to understand environmental crime theoretically and conceptually and as an empirical program of research, to develop better sources of data to understand the universe of environmental crime and its victims.[49] Many have begun to label the study of social harms as zemiology.[50] Taking an "eco-justice" perspective, Rob White frames environmental crime victims in terms who or what is being harmed, including: "Environmental justice -the victim is humans; Ecological justice- the victim is specific environments-; and Species justice – the victim is animals, and plants".[51]

One can take White's typology as a good starting point for building a broader theoretical base to conceptualize an environmental victimology. Matthew Hall notes that scholars also have increasingly moved toward a more culturally informed approach to victimology to explain the marginalization of environmental crime victims. Such a victimology is "characterized by the recognition of wider forms of harm, suffering and trauma coupled with the growing acceptance that victimization often constitutes a dynamic, evolving process" rather than a singular event.[52] This approach also considers that an environmental victimology focuses as much on environmental harm, rather than environmental crime, given the way such crimes are traditionally marginalized within the criminal justice system. In this, an environmental victimology must consider the cultural dimensions that label certain crimes, such as street crimes as bad, whereas many cultures still delegitimize the idea that environmental crimes cause similarly serious harm to humans, non-human animals, and the natural environment that go legally, politically, and culturally unacknowledged. As Rob White has argued here, defining victims of crime is never a socially neutral enterprise.[53]

Matthew Hall notes that, such a cultural victimology might contain two primary aspects including, "the wider sharing and reflection of individual and collective victimization experiences on the one hand and, on the other, the mapping of those experiences through the criminal justice process…through both these mechanisms, wider definitions of "victimization" become culturally embedded to the point where policy makers are prompted to respond with more targeted reform".[54] A cultural victimology must broader the definition and discussion of harm to include non-human animals and the natural environment.[55] This approach may emphasize the value of making victims of environmental harm whole, through restorative justice mechanisms, such as victim-offender mediation and problem-solving courts.[56] Moving towards a restorative justice model for environmental offenders and their victims provides an active role for victims in the criminal justice process, structural opportunities for problem solving, and the potential for healing social bonds broken in communities due to environmental crime.[57]

Because victimization often results from pollution that acts on bodies as a silent force and criminal punishment of offenders is rare, the empirical study of environmental victimization is stymied by a lack of systematically available data.[58] Outside of a number of studies in the 1990s and early 2000s on environmental justice, recent work on victimization has moved beyond theory development to empirical analysis.[59] Data limitations still make broad and systematic generalizations about environmental victimization difficult both in the United States and globally. Developing datasets on environments and on environmental victims can thus take a broader social harm perspective and incorporate a number of factors into defining and quantifying victimization, as can using government data to approach the same issues from a more limited, legal framework that may also have the power of identifying victims and offenders formally within the legal system to advance the empirical and theoretical study of victims of environmental crime.[60]

Our own work on environmental victimization took the latter route, by exploring victimization in the context of environmental crime prosecutions. This approach has allowed us to develop long-term datasets and build both empirical and theoretical insights into the role of victims in the criminal investigations and prosecutions, to give a sense of who is victimized and why, and how offenders are punished. Our work in this area grew out of community engagement in environmental justice work in Corpus Christi, Texas.[61] From there, we had the opportunity to participate in a federal environmental crime prosecution that leads to the case of *United States v. CITGO*, and the recognition of members of an environmental justice community as victims of air pollution crimes committed by a corporation under the federal Crime Victims' Rights Act.[62] These practical experiences working on the environmental policy process, the criminal enforcement process, and with victims, spurred an interest in who victims are and how they are treated within the U.S. regulatory and legal system, as well as practical avenues for reducing harm to victims generally, but particularly within the most vulnerable communities that experience chronic environmental exposure and harm.[63]

Our work on environmental justice and thinking conceptually and practically about how these marginalized communities might be better protected and treated as victims under federal law spurred other investigations into the role of victims in environmental crime prosecutions. In addition to our aforementioned work on environmental justice communities as victims of environmental crime, we explored the victimization of non-human animals under the law and how it was represented in environmental crime prosecutions.[64] We then described possibilities for expanding procedural rights for environmental crime victims as part of the ongoing movement in the U.S. states to enshrine crime victims' rights within state constitutions.[65] We then undertook an initial exploration of the role of environmental crime victims themselves within environmental crime prosecutions.[66] All of these studies lead to our interest in providing a systematic account of the universe of environmental crime victims within the U.S. criminal enforcement regime. We explore the institutionalization of that apparatus below, followed by a discussion of its deterrent value for preventing environmental crimes and protecting victims.

The Evolution of Criminal Enforcement

Mounting evidence of environmental harm by the mid-twentieth century, combined with salient focusing events, such as the Donora Smog, Santa Barbara Oil Spill, and the Cuyahoga River Fires, drew public criticism and prompted Congress to pass a number of important environmental laws in the 1970s.[67] Some of the more important laws include the National Environmental Policy Act (NEPA), CAA, CWA, and RCRA, along with the Toxic Substances Control Act (TSCA),[68] and the Federal Insecticide, Fungicide, and Rodenticide Act (FIFRA).[69]

Around this same time, with the backdrop of these increasingly prevalent environmental problems not being adequately controlled by the U.S. states, President Nixon's Advisory Council on Executive Organization issued the Ash Council Memo, calling for the creation of a single federal environmental agency to tackle these problems.[70] While Nixon demurred, preferring a weaker approach, the president relented and forwarded plans to Congress for an executive reorganization to create EPA.[71] The agency was organized through Executive Order 1110.2.[72] The reorganization merged responsibilities and resources from extant federal agencies, such as the Council on Environmental Quality, Federal Radiation Council, and the Department of Agriculture.[73] The following units were initially established: Administrator, Office of the Administrator, Deputy Administrator, Pesticides Office, Solid Waste Office, Air Pollution Control Office, Water Quality Office, Assistant Manager for Standards and Enforcement and the General Counsel, Assistant Administrator for Planning and Management, Assistant Administrator for Research and Monitoring, and Regional Offices.[74] Much of the enforcement, compliance, and coordination activities would fall to the regional offices.[75]

Unlike previous efforts, the EPA would have a singular missions allowing the new agency to focus its resources exclusively on environmental protection, as well as taking over the lead from the U.S. states.[76] While the founding of EPA represented a historic effort by the federal government to take the lead on environmental regulation, the agency faced serious obstacles to statutory interpretation and enforcement, including a lack of direction provided by Congress, tight deadlines for compliance with Congressional mandates, the lack of tested pollution control technologies, and the steep cost of adopting these new technologies.[77] Despite these limitations, first administrator, William Ruckelshaus, worked hard to create an cohesive enforcement culture and presence with the regulated community that included creation of the infrastructure for developing publicly owned water treatment works (POTWs), the largest investment in public infrastructure at the time after the Federal-Aid Highway Act of 1956 that funded the interstate highway system, instituted a ban on DDT, initiated Operation 5000 that moved to close 5,000 open landfills and convert them to sanitary landfills, and the agency filed a series of high-profile lawsuits against the Cities of Cleveland, Detroit, and Atlanta, for water pollution violations.[78]

Not long after EPA's founding, it became increasingly apparent to the new agency that the host of new environmental laws it was tasked with enforcing would encounter serious and chronic violations of law and would require a new set of tools to both ensure compliance and deter criminal offenses.[79] A number of other countries were wrestling with the same problem and were moving to enhance enforcement tools for the environment.[80] Yet, stronger penalties and the infrastructure to enforce the law and deter serious and chronic violations would be forthcoming for more than half a century. Misdemeanor penalties for environmental violations made their way into federal law in the United States via the Rivers and Harbors Act and Lacey Act, passed at the turn of the twentieth century.[81] It was not until the 1980s that Congress moved to add criminal provisions to environmental laws, including felonies, to punish a variety of polluting behaviors.[82] Spurred by the "get tough on crime" movement to enhance federal penalties for a variety of crimes (and the formation of and suggestions provided by the United States Sentencing Commission), sufficient support for enhancing penalties for environmental offenses warranted Congressional action, which upgraded penalties to RCRA in 1984, the CWA in 1987, and the CAA in 1990.[83]

The process of institutionalizing resources to pursue criminal investigations against environmental offenders might be traced to early efforts such as the EPA's Major Source Enforcement Efforts (MSEE) that attempted to bring facilities into compliance with the CWA and CAA; it uncovered widespread non-compliance.[84] By 1981, the agency organized the Office of Enforcement and hired 22 criminal investigators as well as created further organizational ties to the DOJ when DOJ attorney Peter Beeson was tapped as Director of the Office.[85] Further investments in criminal enforcement came with the passage of the Medical Waste Tracking Act in 1988, providing EPA criminal investigators full law enforcement authority; the following year they were approved by the U.S. Attorney General to carry firearms in their official capacities.[86] The Pollution Prosecution Act was authorized in 1990, allowing EPA to hire up to 200 criminal enforcement staff that are currently housed within the Criminal Investigation Division (EPA-CID).[87]

The Public Lands Division was established in 1909, tasking DOJ to oversee natural resources law.[88] From that time to the middle part of the twentieth century, the approach taken by the federal government to enforce violations of law has generally included seeking civil sanctions or injunctive relief, with criminal cases undertaken by DOJ for pollution activities uncommon.[89] Institutionalization of resources for criminal prosecution within DOJ began in the 1970s with the creation of the Environmental Crimes Unit.[90] In 1982, the Environmental Crimes Section (DOJ-ECS) was organized to institutionalize resources for criminal prosecution and it became its own organizational unit in 1987, located within the Environment and Natural Resources Division (ENRD).[91] Currently, 43 attorneys and a dozen staff are housed within DOJ-ECS and specialize in the prosecution of environmental crimes.[92]

When violations of law involve serious harm or culpable conduct, EPA may choose to open a criminal investigation and seek criminal sanctions for the violator, moving the approach from one of compliance with the law, to punishment and deterrence of crime.[93] In such cases, EPA collaborates with DOJ to pursue criminal prosecution of the offender.[94] Environmental crimes in this context are defined by EPA as simply a "knowing or negligent violation of an environmental law".[95] An appropriate state environmental agency may also pursue criminal prosecution of environmental offenders both exclusively and in addition to actions undertaken by federal prosecutors.[96]

There are, according to EPA, seven common signs of an environmental crime including, "[s]trong, offensive, or unusual chemical odors; [l]arge numbers of dead birds, fish or other animals; [p]ipes or valves that bypass waste treatment systems; [t]ank trucks discharging into drains, manholes or surface waters; [o]ily slicks on bodies of water; [c]orroded, leaking waste containers; and [d]rums or containers dumped at odd hours in out-of-the-way places."[97] When investigating potential crimes, EPA, DOJ, and state and local law enforcement agencies often collaborate with one another and share resources.[98] In the criminal enforcement scheme, EPA monitors and polices violations and prosecutors from the DOJ work with EPA-CID investigators to decide to prosecute violators. This relationship may have its origins in 1976 when a joint hazardous waste force initiated 52 RCRA civil actions.[99] Criminal investigators possess a sizable amount of professional discretion when pursuing investigations.[100] They build their cases from information provided by whistleblowers, information uncovered during civil inspections, or evidence provided through required regulatory disclosures. When they have sufficient evidence of wrongdoing, they work with DOJ prosecutors to begin criminal proceedings by either convening a grand jury in the hopes of returning an indictment or presenting a criminal complaint to a federal judge to determine if there is probable cause to issue an arrest warrant.[101]

Conclusion

In this chapter, we described the theories of compliance and deterrence in environmental crime enforcement, theoretical frameworks for environmental crime victimization, and the evolution of criminal enforcement tools for the environment in the United States. From a social harm perspective, thinking of victimization more broadly suggests that environmental crime victims are numerous in society and generally ignored by the mass media only in exceptional circumstances. The criminal enforcement regime was developed in the early 1980s in the United States and a number of other countries as they realized that widespread noncompliance with the law had and would continue to exist, absent the imposition of a criminal enforcement scheme to better monitor serious and chronic violations of law and to exert the full force of law to punish particular offenders and to set larger, generalized deterrence examples for the regulated community as a whole.

Criminal enforcement was always meant to be a strategic and limited application of the law that would work alongside both civil enforcement and state efforts to ensure compliance with the law and to deter criminal acts. Whether this system truly deters environmental crimes and protects victims is hard to assess and the deeper empirical question of deterrence may ultimately prove exceptionally hard to pin down. Since our only official sources of data on victimization come from those cases where victims are acknowledged as a part of criminal prosecutions, we have limited, systematic data on the subject that has stymied empirical research in the area. In the following chapters, we tackle this question as systematically as we can, to provide broader insight into the universe of victims within criminal prosecutions and the punishments meted out to offenders. While our work will not be the definitive piece on this universe or the deterrent value of the criminal enforcement regime, it will be the first systematic attempt to explore these questions since the criminal enforcement system was institutionalized in the United States in the early 1980s.

Notes and References

1. U.S. EPA, 2024, Basic Information on Enforcement, Available from:
https://www.epa.gov/enforcement/basic-information-enforcement.
2. Michael L. Rustad, Thomas H. Koenig, and Erica R. Ferreira, 2013, Restorative Justice to Supplement Deterrence-Based Punishment: An Empirical Study and Theoretical Reconceptualization of the EPA's Power Plant Enforcement Initiative, 2000–2011, 65 Oklahoma Law Review, 427.
3. Michael J. Lynch, 2017, The Sentencing/Punishment of Federal Environmental/Green Offenders, 2000–2013, 38 Deviant Behavior, 991–995.
4. U.S. EPA, 2017, OCEFT At a Glance, Available from: https://www.epa.gov/sites/default/files/2019-05/documents/oceft-at-a-glance-aug2017.pdf.
Earl E. Devaney, 1994, The Exercise of Investigative Discretion, 3–4, Available from: https://www.epa.gov/sites/production/files/documents/exercise.pdf.
U.S. EPA, 2024, Criminal Enforcement Overview, Available from: https://www.epa.gov/enforcement/criminal-enforcement-overview.
5. Gary Becker, 1968, Punishment: An Economic Approach, 169 J. of Pol. Econ. 183.
6. Sahramaki, Lina, Lars Korsell, and Tehri Kankaanrata, 2015, Prevention of Environmental Crime through Enforcement—Finland and Sweden Compared, 16 Journal of Scandinavian Studies in Criminology and Crime Prevention, 41–59.
7. U.S. EPA, 2024, Criminal Provisions of the Resource Conservation and Recovery Act, Available from: https://www.epa.gov/enforcement/criminal-provisions-resource-conservation-and-recovery-act-rcra#seven.
8. Raymond W. Mushal, 2009, Up From the Sewers: A Perspective on the Evolution of the Federal Environmental Crimes Program, Utah Law Review, 1103–1107.
9. Richard J. Lazarus, 1994, Assimilating Environmental Protection into Legal Rules and the Problem with Environmental Crime, 27 Loyola Law Review, 867–870.

10. Sidney M. Wolf, 1993, Finding an Environmental Felon Under the Corporate Veil: The Responsible Corporate Officer Doctrine and RCRA, 9 Journal of Land Use & Environmental Law, 1–2.
11. Robert I. McMurry and Stephen D. Ramsey, 1986, Environmental Crime: The Use of Criminal Sanctions in Enforcing Environmental Laws, 19 Loyola of Los Angeles Law Review, 1157–1160.

    Robert T. McGovern, 1985, United States v. Johnson & Towers, Inc.: Corporate Employee Criminal Liability under RCRA, 2 Pace Environmental Law Review, 316, 328–329.
12. Francois Comte, 2006, Environmental Crime and the Police in Europe: A Panorama and Possible Paths for Future Action, 15 European Energy and Environmental Law Review, 190–231.
13. Ronald M. Broudy, 1992, RCRA and the Responsible Corporate Officer Doctrine: Getting Tough on Corporate Offenders by Sidestepping the Mens Rea Requirements, 80 Kentucky Law Journal, 1055, 1072–1073.
14. U.S. EPA, 2024, U.S. Environmental Protection Agency Criminal Enforcement Program: America's Environmental Crime Fighters. Available from: https://www.epa.gov/sites/production/files/documents/oceftbrochure.pdf;

    Public Employees for Environmental Responsibility (PEER), 2019. EPA CID Agent Count, Available from: https://www.peer.org/wp-content/uploads/2019/11/11_21_19-Federal_Pollution_EPA_CID_Agent_Count.pdf.
15. U.S. EPA, 2024, U.S. Environmental Protection Agency Criminal Enforcement Program: America's Environmental Crime Fighters. Available from: https://www.epa.gov/sites/production/files/documents/oceftbrochure.pdf.
16. Earl E. Devaney, 1994, The Exercise of Investigative Discretion, 3–4, Available from: https://www.epa.gov/sites/production/files/documents/exercise.pdf.
17. EPA and DOJ may pursue both civil and criminal measures and they often apply criminal tools for particularly egregious offenses on top of pursuing civil damages, see: Robert I. McMurry and Stephen D. Ramsey, 1986, Environmental Crime: The Use of Criminal Sanctions in Enforcing Environmental Laws, 19 Loyola of Los Angeles Law Review, 1163–1164.

18. U.S. EPA, 2024, Basic Information on Enforcement, Available from:
    https://www.epa.gov/enforcement/basic-information-enforcement.
19. U.S. EPA, 2021, Using All Appropriate Injunctive Relief Tools in Civil Enforcement Settlements, Available from: https://www.epa.gov/enforcement/using-all-appropriate-injunctive-relief-tools-civil-enforcement-settlements.
20. U.S. EPA, 2024, Guidance on Use of Penalty Policies in Administrative Litigation, Available from: https://www.epa.gov/enforcement/guidance-use-penalty-policies-administrative-litigation.
    U.S. EPA, 2024, Securing Mitigation as Injunctive Relief in Certain Civil Enforcement Settlements, Available from: https://www.epa.gov/enforcement/securing-mitigation-injunctive-relief-certain-civil-enforcement-settlements-2nd-edition.
    U.S. EPA, 2024, Supplemental Environmental Projects (SEPs), Available from: https://www.epa.gov/enforcement/supplemental-environmental-projects-seps.
21. U.S. EPA, 2021, Using All Appropriate Injunctive Relief Tools in Civil Enforcement Settlements, Available from: https://www.epa.gov/enforcement/using-all-appropriate-injunctive-relief-tools-civil-enforcement-settlements.
22. Lucia Ann Silecchia, 2011, Ounces of Prevention and Pounds of Cure: Developing Sound Policies for Environmental Compliance Programs, 7 Fordham Environmental Law Journal, 583.
    For examples of a state approach here, see:
    CalEPA, 2024, Supplemental Environmental Projects, Available from: https://calepa.ca.gov/enforcement/supplemental-environmental-projects/.
23. One can easily browse the Enforcement and Compliance History Online (ECHO) database created by EPA to see that civil enforcement is much more frequent and extensive than criminal enforcement, see:
    U.S. EPA, 2024, Enforcement and Compliance History Online, Available from: https://echo.epa.gov/.
24. Carole M. Billiet and Sandra Rousseau, 2014, How Real Is the Threat of Imprisonment for Environmental Crime? 37 European Journal of Law and Economics, 183–186.

Raymond Paternoster, 2010, How Much Do We Really Know About Criminal Deterrence? 100 Journal of Criminal Law & Criminology at 765–768. Michael J. Lynch,

25. David M. Uhlmann, 2019, Prosecutorial Discretion and Environmental Crime Redux: Charging Trends, Aggravating Factors, and Individual Outcome Data For 2005–2014, 8 Michigan Journal of Environmental & Energy Law, 312, 351–362.
26. Joshua Ozymy, Bryan Menard, and Melissa L. Jarrell, 2021. Persistence or Partisanship: Exploring the Relationship between Presidential Administrations and Criminal Enforcement by the U.S. Environmental Protection Agency, 1983–2019. 81 (1) Public Administration Review, 49–63.
27. Kimberly L. Barrett, Paul B. Stretesky, and Michael Long, 2016, The Weak Probability of Punishment for Environmental Offenses and Deterrence of Environmental Offenders: A Discussion Based on USEPA Criminal Cases, 1983–2013, 37 Deviant Behavior at 1096–1099.
28. Joshua Ozymy and Melissa L. Jarrell, 2021, Does the Criminal Enforcement of Federal Environmental Law Reduce Crime? The Case of the Resource Conservation and Recovery Act. 11 (1) Environmental and Earth Law Journal, 65–88.
29. For empirical studies on state and local environmental crime enforcement, see: Matthew S. Crow, Tara O'Conner Shelley, and Paul B. Stretesky, 2013. Camouflage-Collar Crime: An Examination of Wildlife Crime and Characteristics in Florida, 34 Deviant Behavior, 635.

    Joshua C. Cochran, Michael J. Lynch, Elisa L. Toman, and Ryan T. Shields, 2018. Court Sentencing Patterns for Environmental Crimes: Is there a "Green" Gap in Punishment? 34 Journal of Quantitative Criminology, 37.
30. For some of our work on the role of victims in environmental crime prosecutions that lead up to this broader work herein, see:

    Joshua Ozymy and Melissa L. Jarrell, 2020. Exploring the Role of Victims in Federal Environmental Crime Prosecutions. 57 (1) Willamette Law Review, 101–137.
31. European Commission, 2024, Environmental Crime, Available from: https://commission.europa.eu/strategy-and-policy/policies/justice-and-fundamental-rights/criminal-justice/environmental-crime_en.

32. Ibid.
33. European Parliament of the Council, 2008, Directive 2008/99/EC of the European Parliament and of the Council of 19 November 2008 on the protection of the environmental through criminal law, Available from: https://eur-lex.europa.eu/legal-content/EN/TXT/?uri=CELEX:32008L0099.
34. European Commission, 2024, Environmental Crime, Available from: https://commission.europa.eu/strategy-and-policy/policies/justice-and-fundamental-rights/criminal-justice/environmental-crime_en.
35. Ricardo Pereira, 2015, Environmental Criminal Liability and Enforcement in European and International Law, Queen Mary Studies in International Law Series, BRILL.
36. Ibid.

    For two Scandinavian countries, Finland and Sweden, both countries criminalize a range of environmental behaviors, see:

    Sahramaki, Lina, Lars Korsell, and Tehri Kankaanrata, 2015, Prevention of Environmental Crime Through Enforcement—Finland and Sweden Compared, 16 Journal of Scandinavian Studies in Criminology and Crime Prevention, 41–59.
37. Ibid.
38. Ibid.
39. Rob White, 2015, Environmental Victimology and Ecological Justice, In Wilson D and Ross S (Eds.) Crime, Victims and Policy: International Contexts, Local Experiences. Basingstoke: Palgrave Macmillan, pp. 33–52.

    Rob White, 2018, Green Victimology and Non-Human Victims, 24 International Review of Victimology, 239.
40. Kathleen F. Brickey, 1996, Environmental Crime at the Crossroads: The Intersection of Environmental and Criminal Law Theory, 487 Tulane Law Review, 494–495.

    Ronald G. Burns and Michael J. Lynch, 2004, The Sourcebook on Environmental Crime, LFB Publishers.
41. Matthew Hall, 2013, Victims of Environmental Harm: Rights, Recognition and Redress Under National and International Law, London: Routledge.

    E. Skinnider, 2011, Victims of Environmental Crime—Mapping the Issues. Vancouver: The International Centre for Criminal Law Reform and Justice Policy.

42. Melissa L. Jarrell, 2007, Environmental Crime and the Media: News Coverage of Petroleum Refining Industry Violations, New York: LFB.
43. Paddy Hillyard and Steve Tombs, 2004, Beyond Criminology, In P. Hillyard, C. Pantazis, S. Tombs, and D. Gordon (Eds.) Beyond Criminology: Taking Harm Seriously, Pluto Press.

    Melissa L. Jarrell, Joshua Ozymy, and William Sandel, 2017, Where the Wild Things Are: Animal Victimization at the Intersection of Wildlife Law and Environmental Law, 3 Contemporary Justice Review, 320.
44. Melissa L. Jarrell, 2007, Environmental Crime and the Media: News Coverage of Petroleum Refining Industry Violations, New York, LFB.

    A good definition of "traditional victims of crime" to contrast with the treatment of environmental crime victims in the criminal justice system is provided by Matthew Hall: "'Traditional victims' here refers broadly to all classifications of victims of crime which have received extended attention by criminologists and victimologists, and by policy-makers, in most Hall developed jurisdictions".

    Matthew Hall, 2013, Victims of Environmental Harm: Rights, Recognition and Redress Under National and International Law, Taylor & Francis Group, 3–4.
45. Matthew Hall, 2014, Environmental Harm and Environmental Victims: Scoping Out a 'Green Victimology', 20 International Review of Victimology, 139.

    The effects of various environmental crimes are numerous. For a discussion of these effects, see:

    Joanna Shapland and Matthew Hall, 2007, What Do We Know About the Effects of Crime on Victims? 14 International Review of Victimology, 178–180.
46. Michael J. Lynch, 2006, The Greening of Criminology: A Perspective on the 1990s, London: Routledge.

    Matthew Hall, 2012, Environmental Victims: Challenges for Criminology and Victimology in the 21st Century, 13 Journal of Criminal Justice and Security, 371.

    Rob White, 2011, Transnational Environmental Crime: Toward an Eco-Global Criminology, London: Routledge.

47. Rob White, 2016, Inter-species violence: Humans and the harming of animals, In Stubbs J. and Tomsen S. (Eds.) Australian Violence: Crime, Criminal Justice and Beyond. Sydney: The Federation Press, pp. 176–193.

Melanie Flynn and Matthew Hall, 2017, The Case for a Victimology of Nonhuman Animal Harms 20 Contemporary Justice Review, 299–318.

Joshua Ozymy and Melissa L. Jarrell, 2021, Exploring the Role of Non-Human Animal Victims in Federal Environmental Crime Prosecutions, 27 Animal Law Review, 35–56.

48. Paul B. Stretesky and Michael J. Lynch, 1999, Corporate Environmental Violence and Racism. 30 Crime, Law and Social Change, 163.

Nigel South, Avi Brisman and Piers Bierne, 2013, A Guide to a Green Criminology. In Nigel South and Avi Brisman (Eds.) Routledge International Handbook of Green Criminology, New York: Routledge.

49. Vincenzo Ruggiero, 2013, The Environment and Crimes of the Economy. In Nigel South and Avi Brisman (Eds.) Routledge International Handbook of Green Criminology, New York: Routledge.

50. Matthew Hall, 2014, Environmental Harm and Environmental Victims: Scoping Out a 'Green Victimology', 20 International Review of Victimology, 131.

51. Rob White, 2013, Environmental Harm: An Eco-Justice Perspective, Bristol: Policy Press.

Rob White, 2018, Green Victimology and Non-Human Victims, 24 International Review of Victimology, 240–241.

52. Matthew Hall, 2017, Exploring the Cultural Dimensions of Environmental Victimization, 3 Palgrave Communications, 2.

53. Rob White, 2011, Transnational Environmental Crime: Towards an Eco-global Criminology, Abingdon: Routledge Publishing.

54. Matthew Hall, 2017, Exploring the Cultural Dimensions of Environmental Victimization, 3 Palgrave Communications, 4.

55. Piers Beirne and Nigel South, 2007, Issues in Green Criminology: Confronting Harms Against Environments, Humanity and Other Animals, Willan Publishing: Cullompton.

Rob White, 2008, Crimes Against Nature: Environmental Criminology and Ecological Justice, Willan Publishing: Cullompton.

56. Neil Sipe, 2007, An Empirical Analysis of Environmental Mediation, 64 Journal of the American Planning Association, 275–285.

    Shigeru Matsumoto, 2011, A Duration Analysis of Environmental Alternative Dispute Resolution in Japan, 70 Ecological Economics, 659–666.

    Antony Pemberton, 2014, Environmental Victims and Criminal Justice: Proceed with Caution, In Spapens et al. (Eds.) Environmental Crime and Its Victims. UK: Routledge.

57. Lorenzo Natali, Matthew Hall, Laura Centemeri, Peter J. Burgess, and Sezin Topcu, 2022, A Green Criminological Approach to Environmental Victimization and Reparation: A Case for Environmental Restorative Justice, In Centemeri et al. (Eds.) Re-thinking Post-Disaster Recovery. UK: Taylor & Francis.

    For examples of compensatory schemes for restitution through a criminal process, see:

    Rob White, 2014, Eco-Justice and Problem-Solving Approaches to Environmental Crime and Victimisation, In Spapens et al. (Eds.) Environmental Crime and Its Victims, UK: Routledge.

    Matthew Hall, 2014, Victims of Environmental Crime: Routes for Recognition, Restitution and Redress, In Spapens et al. (Eds.) Environmental Crime and Its Victims, UK: Routledge.

58. Melissa L. Jarrell and Joshua Ozymy, 2014. Few and Far Between: Understanding the Role of the Victim in Federal Environmental Crime Cases, 61 Crime, Law, and Social Change, 563–584.

59. Michael J. Lynch 1990, The Greening of Criminology: A Perspective on the 1990s. Criminologist, 11–12.

    Michael J. Lynch and Paul B. Stretesky, 2003, The Meaning of Green: Contrasting Criminological Perspectives, 7 Theoretical Criminology, 217.

    Penny Green, Tony Ward, and Kristen McConachie, 2007, Logging and Legality: Environmental Crime, Civil Society, and the State, 34 Social Justice, 94-110.

    Michael J. Lynch, 2017, The Sentencing/Punishment of Federal Environmental/Green Criminal Offenders, 2000-2013, 38 Deviant Behavior, 1008.

Paul B. Stretesky, Michael J. Lynch, Michael A. Long, & Kimberly L. Barrett, 2017, Does the Modernization of Environmental Enforcement Reduce Toxic Emissions? An Examination of Self-Policing, Criminal Prosecutions, and Toxic Releases in the United States, 1988-2014, 37 Sociological Spectrum, 48.

60. Carole Gibbs and Sally S. Simpson, 2009, Measuring Corporate Environmental Crime Rates: Progress and Problems, 51 Crime, Law, and Social Change, 87–107.

Michael J. Lynch, Paul B. Stretesky, and Ronald G. Burns, 2004, Slippery Business: Race, Class, and Legal Determinants of Penalties Against Petroleum Refineries, 34 Journal of Black Studies, 421–440.

61. Joshua Ozymy and Melissa Jarrell, 2015, Corporate Environmental Crime and Environmental Victimization: Exploring New Legal Precedents for Securing Recognition and Restitution for Environmental Justice Communities, 8 Environmental Justice, 47–50.

62. Melissa L. Jarrell and Joshua Ozymy, 2012, Real Crime, Real Victims: Environmental Crime Victims and the Crime Victim's Rights Act (CVRA), 58 Crime, Law, and Social Change, 373–389.

Crime Victims' Rights Act: (18 U.S.C. § 3771, 1984).

Joshua Ozymy and Melissa L. Jarrell, Exploring Prospects for Environmental Justice as the EPA hits the Half-Century Mark, 47 (1) Ecology Law Quarterly Currents, 269–280.

63. Suzie Canales, Joshua Ozymy, and Melissa L. Jarrell, 2012, Risk Assessment or Risk Acceptance: Why the EPA's Attempts to Achieve Environmental Justice Have Failed and What They Can Do About It, 5 Environmental Justice, 59–62.

Joshua Ozymy and Melissa L. Jarrell, 2017, Implementing a Partial Buyout of an Environmental Justice Community, Environmental Justice, 43-50.

Joshua Ozymy and Melissa Jarrell, 2019, Righting and "Writing" Wrongs: A Postmortem on a Decade of Environmental Justice Activism in Corpus Christi, Texas, 11 Environmental Justice, 23–31.

64. Joshua Ozymy and Melissa L. Jarrell, 2021, Exploring the Role of Non-Human Animal Victims in Federal Environmental Crime Prosecutions, 27 Animal Law Review, 35–56.

Melissa L. Jarrell, Joshua Ozymy, and William Sandel, 2017, Where the Wild Things Are: Animal Victimization at the Intersection of Wildlife Law and Environmental Law, 3 Contemporary Justice Review, 319–335.

65. Joshua Ozymy and Melissa L. Jarrell, 2020, Of Sex Crimes and Fencelines: How Recognition of Environmental Justice Communities as Crime Victims under State and Federal Law can help Secure Environmental Justice, 38 (1) Pace Environmental Law Review, 109–146.
66. Joshua Ozymy and Melissa L. Jarrell, 2020, Exploring the Role of Victims in Federal Environmental Crime Prosecutions, 57 (1) Willamette Law Review, 101–137.
67. Richard Nixon, 1970, Special Message from the President to the Congress About Reorganization Plans to Establish the Environmental Protection Agency and the National Oceanic and Atmospheric Administration, Available from: https://www.epa.gov/archive/epa/aboutepa/reorganization-plan-no-3-1970.html.

National Park Service, 2022, The 1969 Cuyahoga River Fire, Available from: https://www.nps.gov/articles/story-of-the-fire.htm.

Air Quality Alliance, 2022, The Beast of 1943: A History of Smog in Los Angeles, Available from: https://airqualityalliance.com/2020/07/26/the-beast-of-1943-a-history-of-smog-in-los-angeles/.

National Oceanic and Atmospheric Administration, 2022, 45 Years After the Santa Barbara Oil Spill, Looking at a Historic Disaster Through Technology, Available from: https://response.restoration.noaa.gov/about/media/45-years-after-santa-barbara-oil-spill-looking-historic-disaster-through-technology.html.

68. Toxic Substances Control Act, 53 U.S.C. §2601 (1976). Authorizes EPA to regulate chemical substances, from manufacturing, use in commerce, and importation. U.S. EPA, 2024, Learn About the Toxic Substances Control Act, Available from: https://www.epa.gov/assessing-and-managing-chemicals-under-tsca/learn-about-toxic-substances-control-act-tsca.
69. Federal Insecticide, Fungicide, and Rodenticide Act (FIFRA), 7 U.S.C. §136 (1972). Authorizes EPA to manage health risks from pesticides and to regulate commercial applicators, importation of pesticides, and other issues related to pesticides.

U.S. EPA, 2024, Summary of the Federal Insecticide, Fungicide, and Rodenticide Act, Available from: https://www.epa.gov/laws-regulations/summary-federal-insecticide-fungicide-and-rodenticide-act.

70. Ash Council Memo, 1970, Federal Organization of Environmental Protection, Available from: https://www.epa.gov/archive/epa/aboutepa/ash-council-memo.html.
71. EPA Historical Publications, 1992, The Guardian: Origins of the EPA, Available from: https://www.epa.gov/archive/epa/aboutepa/guardian-origins-epa.html.

    The Council noted a number of options, but also acknowledged the enormity of the problem and serious need to have a stronger regulatory agency over the environment, see:

    U.S. EPA, 2024, Ash Council Memo, Available from: https://www.epa.gov/archive/epa/aboutepa/ash-council-memo.html.
72. Dennis C. Williams, 1993, Why Are Our Regional Offices and Labs Located Where They Are? A Historical Perspective on Siting, Available from: https://www.epa.gov/history/why-are-our-regional-offices-and-labs-located-where-they-are-historical-perspective-siting.
73. U.S. EPA, 2024, EPA's Origin's: Duties Transferred to EPA from Other Federal Agencies, Available from: https://www.epa.gov/archive/epa/aboutepa/epas-origins-duties-transferred-epa-other-federal-agencies.html.
74. Dennis C. Williams, 1993, Why Are Our Regional Offices and Labs Located Where They Are? A Historical Perspective on Siting, Available from: https://www.epa.gov/history/why-are-our-regional-offices-and-labs-located-where-they-are-historical-perspective-siting.
75. U.S. EPA, 2024, Regional and Geographic Offices, Available from: https://www.epa.gov/aboutepa/regional-and-geographic-offices.
76. William D. Ruckelshaus, 1970, EPA's First Administrator on the Establishment of the EPA, Available from: https://www.epa.gov/archive/epa/aboutepa/epas-first-administrator-establishment-epa.html.
77. Political pressure pushed the reorganization forward, despite a lukewarm reception from the Nixon White House, which preferred weaker, diffuse political control over regulation.

Jack Lewis, 1985, The Birth of the EPA, EPA Journal, Available from: https://www.epa.gov/archive/epa/aboutepa/birth-epa.html.

78. Phil Wisman, 1985, EPA History (1970–1985), Available from: https://www.epa.gov/archive/epa/aboutepa/epa-history-1970-1985.html#ruckelshaus.

U.S. EPA, 1972, DDT Ban Takes Effect, Available from: https://www.epa.gov/archive/epa/aboutepa/ddt-ban-takes-effect.html.

New Jersey Department of Environmental Protection, 2022, Overview of POTWs and Discharges to Them, Available from: https://www.nj.gov/dep/dwq/sius.htm.

William D. Ruckelshaus,1988, Environmental Regulation: The Early Days at EPA, EPA Journal, Available from: https://www.epa.gov/archive/epa/aboutepa/environmental-regulation-early-days-epa.html.

Federal-Aid Highway Act of 1956 (P.L. 84–627, 1956).

79. Richard J. Lazarus, 1994, Assimilating Environmental Protection into Legal Rules and the Problem with Environmental Crime, 27 Loyola Law Review, 867–870.
80. Michael R. Pendleton, 1997, Beyond the Threshold: The Criminalization of Logging, 10 Society & Natural Resources, 191–193.
81. U.S. Department of Justice Environmental Crimes Section, 2015, Historical Development of Environmental Criminal Law, Available from: https://www.justice.gov/enrd/about-division/historical-development-environmental-criminal-law.
82. Prosecuting corporate officers was particularly difficult prior to the addition of criminal provisions. An evolution occurred that forced companies to consider who was liable for polluting activities within their organization. See:

David T. Barton, 1991, Corporate Officer Liability Under RCRA: Stringent but not Strict. 4 (3) Brigham Young University Law Review, 1548–1550.

83. U.S. Department of Justice Environmental Crimes Section, 2015, Historical Development of Environmental Criminal Law, Available from: https://www.justice.gov/enrd/about-division/historical-development-environmental-criminal-law.

United States Sentencing Commission, 2023, About the Commission, Available from: https://www.ussc.gov/.

84. Congress was also exerting pressure on EPA at the time to generate significant enforcement outcomes, see:

    Robert I. McMurry and Stephen D. Ramsey, 1986, Environmental Crime: The Use of Criminal Sanctions in Enforcing Environmental Laws, 19 Loyola of Los Angeles Law Review, 1136–1141.
85. Robert I. McMurry and Stephen D. Ramsey, 1986, Environmental Crime: The Use of Criminal Sanctions in Enforcing Environmental Laws, 19 Loyola of Los Angeles Law Review, 1136–1141.

    Employed 23 criminal investigators by 1983 and by 1990, the criminal enforcement program employed 110 people, including 43 investigators (also known as Special Agents),

    Richard J. Lazarus, 1994, Assimilating Environmental Protection into Legal Rules and the Problem with Environmental Crime, 27 Loyola Law Review, 867–870.
86. Raymond W. Mushal, 2009, Up From the Sewers: A Perspective on the Evolution of the Federal Environmental Crimes Program, Utah Law Review, 1103–1107.

    Medical Waste Tracking Act of 1988 (P.L. 100-582).

    John Peter Suarez, 2003, Management Review of the Office of Criminal Enforcement, Forensics and Training, 7, Available from: https://www.epa.gov/sites/production/files/documents/oceft-review03.pdf.
87. The Pollution Prosecution Act of 1990 (P.L. 101-593). The number of criminal investigators varies by source. Estimates range from 145 to around 200, depending on whether one includes support staff,

    U.S. EPA, 2024, U.S. Environmental Protection Agency Criminal Enforcement Program: America's Environmental Crime Fighters. Available from: https://www.epa.gov/sites/production/files/documents/oceftbrochure.pdf;

    Public Employees for Environmental Responsibility (PEER), 2019. EPA CID Agent Count, Available from: https://www.peer.org/wp-content/uploads/2019/11/11_21_19-Federal_Pollution_EPA_CID_Agent_Count.pdf.
88. DOJ-ENRD, 2019, History, Available from: https://www.justice.gov/enrd/history.

89. Robert I. McMurry and Stephen D. Ramsey, 1986, Environmental Crime: The Use of Criminal Sanctions in Enforcing Environmental Laws, 19 Loyola of Los Angeles Law Review, 1136–1141.
90. Richard J. Lazarus, 1994, Assimilating Environmental Protection into Legal Rules and the Problem with Environmental Crime, 27 Loyola Law Review, 867–870.
91. U.S. Department of Justice Environmental Crimes Section, 2015, Historical Development of Environmental Criminal Law, Available from: https://www.justice.gov/enrd/about-division/historical-development-environmental-criminal-law.
92. DOJ-ENRD, 2024, Environmental Enforcement Section (EES): An Overview of Our Practice, Available from: https://www.justice.gov/enrd/overview-our-practice; DOJ-ECS, 2021, Environmental Crimes Section, Available from: https://www.justice.gov/enrd/environmental-crimes-section.
93. Michael J. Lynch, 2017, The Sentencing/Punishment of Federal Environmental/Green Offenders, 2000–2013, 38 Deviant Behavior, 991–995.
94. Earl E. Devaney, 1994, The Exercise of Investigative Discretion, 3–4, Available from: https://www.epa.gov/sites/production/files/documents/exercise.pdf.
95. U.S. EPA, 2017, OCEFT At a Glance, Available from: https://www.epa.gov/sites/default/files/2019-05/documents/oceft-at-a-glance-aug2017.pdf.
96. Lucia Ann Silecchia, 2011, Ounces of Prevention and Pounds of Cure: Developing Sound Policies for Environmental Compliance Programs, 7 Fordham Environmental Law Journal, 583, 595.
97. U.S. EPA, 2017, OECFT At a Glance, Available from: https://www.epa.gov/sites/default/files/2019-05/documents/oceft-at-a-glance-aug2017.pdf.
98. Joel A. Mintz, 2006, Some Thoughts on the Interdisciplinary Aspects of Environmental Enforcement, 36 Environmental Law Reporter, 10,495–10,496,
99. Robert I. McMurry and Stephen D. Ramsey, 1986, Environmental Crime: The Use of Criminal Sanctions in Enforcing Environmental Laws, 19 Loyola of Los Angeles Law Review, 1136–1141.

100. Robert I. McMurry and Stephen D. Ramsey, 1986, Environmental Crime: The Use of Criminal Sanctions in Enforcing Environmental Laws, 19 Loyola of Los Angeles Law Review, 1161.

Congress was not strict in deciding which environmental violations result in civil, administrative, or criminal enforcement, opening up prosecutorial discretion on how to proceed in particular cases,

David M. Uhlmann, 2014, Prosecutorial Discretion and Environmental Crime, 38 Harvard Environmental Law Review, 159.

101. Joel A. Mintz, 2006, Some Thoughts on the Interdisciplinary Aspects of Environmental Enforcement, 36 Environmental Law Reporter, 10,496–10,497.

According to EPA: "federal law enforcement agents - with full Federal authority to conduct investigations, carry firearms, make arrests, and execute search and arrest warrants", "investigate serious environmental crimes", as do "[s]pecially trained investigators, chemists, engineers, technicians... and [a]ttorneys with environmental crimes expertise... EPA special agents talk and listen to suspects and witnesses, conduct surveillance, seize and analyze records, find people and information, work with forensics experts, prosecutors and other police involved, analyze evidence and data and testify in court." see: U.S. EPA, 2017, OECFT At a Glance, at 19, Available from: https://www.epa.gov/sites/default/files/2019-05/documents/oceft-at-a-glance-aug2017.pdf.

CHAPTER 3

Employees as Victims

Abstract The goal of this chapter is to explore the role of employees as victims of environmental crime in the United States. Employees are the largest single group of victims under consideration in our analysis in the manuscript. Out of 117 total prosecutions, 51 prosecutions or about 44% of overall prosecutions involve employees as environmental crime victims.

Keywords Employees · Victims · Victimization at work

Matthew Lawrence Bowman, the former president of Port Arthur Chemical and Environmental Services (PACES) and CES Environmental Services located near Houston, Texas, oversaw companies that produced caustic materials for paper mills.[1] Bowman oversaw the illegal transportation and disposal of hazardous waste and knowingly failed to protect his employees. On two occasions, December 18, 2008 and April 14, 2009, truck drivers were exposed to hydrogen sulfide, a poisonous gas, and died as the result of their injuries.[2] Bowman was charged with giving false statements and violations of the Occupational Safety and Health Act.[3] He was sentenced to serve 12 months of incarceration and to pay a $5,000 fine.[4]

J. Ozymy and M. Jarrell Ozymy, *The U.S. Administrative State and the Protection of Environmental Crime Victims*, Palgrave Studies in Green Criminology, https://doi.org/10.1007/978-3-031-62680-7_3

The prosecution of Matthew Lawrence Bowman paints a picture of the risk that employees face every day in the workplace that is exacerbated when company owners and managers willful or negligently act without their regard. The case also illustrates the collaborative nature of environmental crime investigations, where EPA-CID, DOJ-ECS, Texas Commission on Environmental Quality (TCEQ), OSHA, and numerous other agencies worked together to hold an individual accountable for his intentional and systematic actions that lead to the death of his employees.[5] Our analysis of almost three thousand prosecutions shows that few result in identifiable victims at work and even fewer cases involve the prosecution of managers or owners for culpable conduct that lead to their victimization. The prosecution and incarceration of Bowman is thus an example and exception to the rule.

We begin the chapter by describing patterns in prosecutions over time and space, followed by a discussion of charging patterns in prosecutions and patterns in criminal behavior and outcomes of prosecutions.[6] The second section explores the broader themes in prosecutions we uncovered in our analysis of employees as victims. The final section illustrates sentencing and punishment patterns for offenders in these cases, including a discussion of significant prosecutions that influence the broader outcomes from the first section of the chapter. Herein, we discuss all of these prosecutions involving employees as victims to illustrate patterns, themes, and sentencing outcomes and to bring context to these trends for employees and from the analysis of the previous chapter.

Patterns in Prosecutions

We begin exploring the role of employees as victims in Fig. 3.1, which outlines the number of prosecutions with employees as environmental crime victims by EPA fiscal year, 1983–2022. We do not see a prosecution in the data with an employee as an identifiable victim of environmental crime until 1995, when two cases emerge from the data. We find one case in 1997 and one in 1998. This pattern persists with no more than three prosecutions occurring in a given year until 2017, when eight prosecutions emerge. We find seven prosecutions in 2022. A grand total of 51 prosecutions occur in the data since 1983, with a yearly average of 1.3 prosecutions.

In Fig. 3.2, we explore the geography of environmental crime victims, by describing the number of prosecutions with employees as victims of

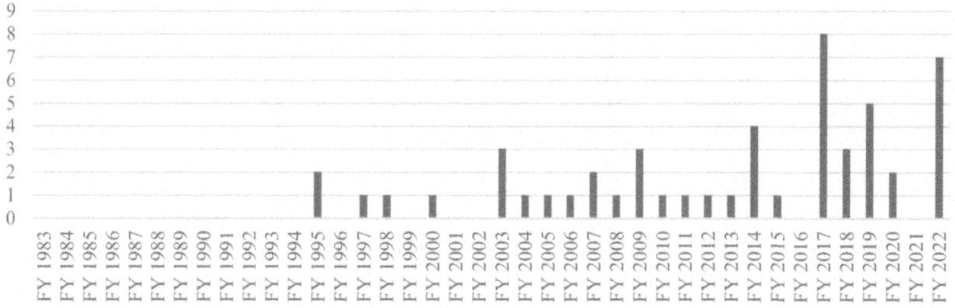

Fig. 3.1 Number of prosecutions with employees as environmental crime victims by EPA fiscal year, 1983–2022 (*Source* EPA Summary of Criminal Prosecutions Database)

environmental crimes by U.S. state, 1983–2022. California dominates the analysis, where 10 prosecutions have occurred since 1983 or 21% of the prosecutions in our data. We find that seven prosecutions were adjudicated in Louisiana, or 15% of the data in our analysis. Six prosecutions were adjudicated in Texas. Generally, all 48 cases in our data are distributed across 20 U.S. states, with California, Louisiana, and Texas making up about 45% of the cases in our analysis.

In Fig. 3.3, we show charging patterns in prosecutions with employees as environmental crime victims, 1983–2022. We provide an overview of major statutes violated here to give a sense of the violations that lead

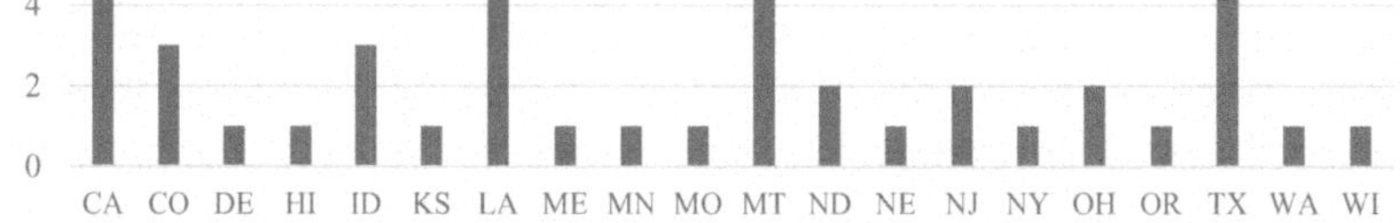

Fig. 3.2 Number of prosecutions with employees as environmental crime victims by U.S. State, 1983–2022 (*Source* EPA Summary of Criminal Prosecutions Database)

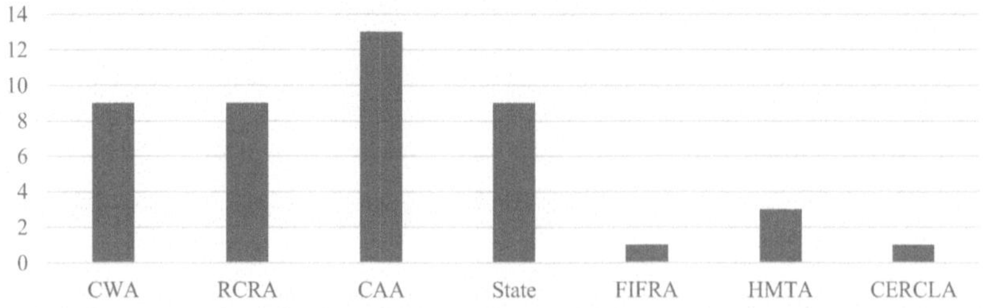

Fig. 3.3 Charging patterns in prosecutions with employees as environmental crime victims, 1983–2022 (*Source* EPA Summary of Criminal Prosecutions Database)

to victimization. Violations of the CAA resulted in 13 prosecutions with employees as victims of environmental crime. We show that in nine prosecutions, at least one defendant was charged for RCRA violations. In nine prosecutions, at least one defendant was charged with a water pollution crime under the CWA, and in nine prosecutions, at least one defendant was charged with violations of state environmental laws.

In Fig. 3.4, we illustrate patterns in criminal behavior occurring in environmental crime prosecutions, where employees were victims of an environmental crime. In addition to air pollution, water pollution, hazardous waste, chemicals, pesticide crimes, and other crimes, we show that in 10 prosecutions or 21% of our data, at least one defendant was charged with making false statements on official documents or reports or to government officials. In four prosecutions or 8% of our data, defendants were charged with conspiracy. In one prosecution, defendants were charged with fraud.

In Fig. 3.5, we describe outcomes of prosecutions where employees are environmental crimes victims. The most common outcome for employees in our data was that at least one employee was injured at work, which occurred in 22 prosecutions or 43% of our data in this chapter. In eight prosecutions or about 16% of our data, at least one employee died as the result of an environmental crime related to the workplace. In 14 prosecutions or 27% of our cases, death and injury resulted from an environmental crime. In seven prosecutions or 14% of cases, an evacuation and injury occurred as the result of an environmental crime. An example of an evacuation occurred during the prosecution of Albert David Hajduk,

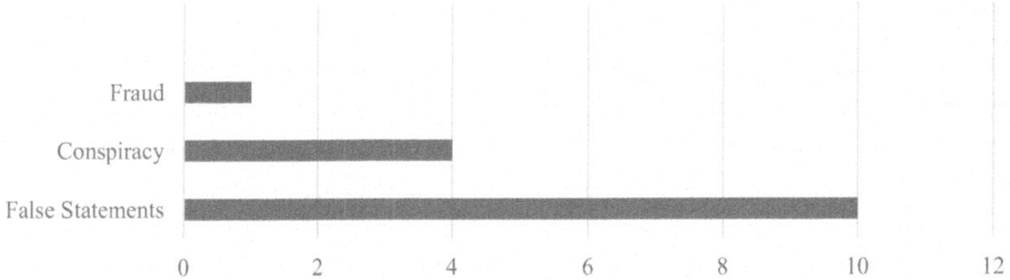

Fig. 3.4 Patterns in criminal behavior for prosecutions with employees as environmental crime victims, 1983–2022 (*Source* EPA Summary of Criminal Prosecutions Database)

who was the operations manager for Luxury Wheels, near Grand Junction, Colorado. The defendant ordered employees to illegally discharge plating waste, submitted falsified discharge monitoring reports, and illegally stored reactive waste acid, causing the fire department to evacuate the facility. The defendant was charged for violations of the CWA and RCRA and was sentenced to five months of incarceration, five months of home detention, 36 months of probation, and a $125 special assessment.[7]

Three cases examples of employees that were killed as consequence of an environmental crime include the prosecution of Lancaster Synthesis,

Fig. 3.5 Categories of victimization for employees that were victims of environmental crime (*Source* EPA Summary of Criminal Prosecutions Database)

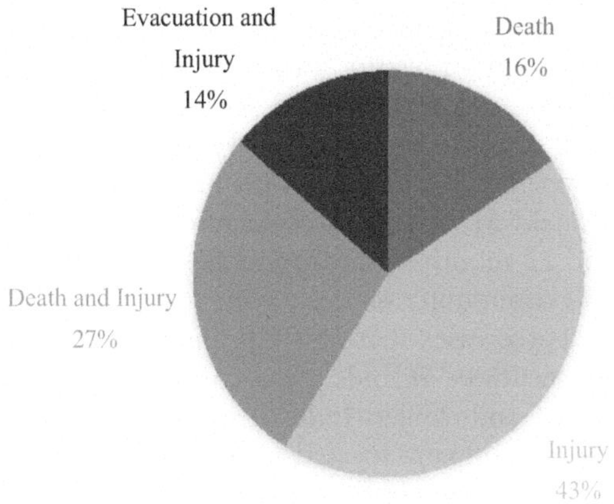

Honeywell International, and Durbin Hartell. Lancaster Synthesis operated a chemical facility in Blythewood, South Carolina, where it manufactured chemicals, including sodium azide. The company closed the facility and in 1993 shipped a drum containing explosive sodium azide without a proper manifest to a hazardous waste disposal company in Cincinnati, Ohio. During transport of the drum at the facility, the drum exploded and the employee transporting the drum died the next day.[8] The company was charged with knowingly transporting hazardous waste without a manifest in violation of RCRA and giving false statements and was sentenced to serve 60 months of probation, pay a $400 special assessment, and provide $250,000 in restitution to the victim's family.[9]

In a similar case of a workplace death to Lancaster Synthesis, a worker died at Honeywell International's plant in Baton Rouge, Louisiana, after 1,800 pounds of antimony chloride were released, striking and killing a worker.[10] The tank had been mislabeled as a non-toxic refrigerant at the company's El Segundo, California plant before being shipped to its plant in Louisiana. Honeywell was charged with violations of the CAA and sentenced to pay a $8 million criminal fine, $2 million to the victim's family, and $2 million in other payments for a total penalty of $12 million.[11]

Durbin Hartell was the plant manager for Spencer Environmental in Portland, Oregon, a used oil and wastewater disposal company. Hartell was charged with violations of the CWA in connection with the prosecution of Spencer Environmental, after a worker died after cleaning a tank with a pressure washer at the company's facility. He was sentenced to pay a $1,250 fine and serve 12 months of probation.[12] In a related case, Donald M. Spencer, president of Spencer Environmental, and the company were charged with hazardous waste violations under RCRA.[13] Donald M. Spencer was sentenced to serve six months of incarceration and 12 months of probation and the company to make $150,000 in fines and community service payments.[14]

Three cases that illustrate injuries to workers in environmental crime prosecutions include prosecution of Norman Teltow, MFA Incorporated, and Miller Industries. Norman Teltow, owner of Gold Metal Paint Company located near Watkins, Colorado, pleaded guilty to illegally treating hazardous waste at the company's facility.[15] Teltow ordered company employees to pour hazardous waste on the floor and engage in other risky practices that left them with nausea, headaches, and dizziness after exposing them to waste containing Methyl Chloride. The defendant was sentenced to 24 months of probation and to pay a $10,000 fine.[16]

MFA Incorporated operated a facility in Martinsburg, Missouri. An employee was loading anhydrous ammonia and the connection between the tanks separated, resulting in severe burns to the employee. The company was charged with violations of the CAA, noting a lack of inspections and replacements for parts caused the connection failure.[17] MFA was sentenced to pay $604,000 in judicially mandated fines, a $100,000 federal fine, and a $125 special assessment.[18] Miller Industries was prosecuted after workers were illegally exposed to hazardous substances that caused chemical burns. The company failed to pay for proper removal and disposal of waste at a closed textile mill building in Lewiston, Maine. The company was charged for violations of RCRA and was sentenced to pay $75,000 in federal fines, a $400 special assessment, $75,000 in restitution to the Maine Hazardous Waste Fund, and serve 12 months of probation.[19]

Themes in Prosecutions

We now move in Fig. 3.6 to provide a typology of employees as environmental crime victims. We explore each case and use our best judgment to ascertain the primary crime at the center of the prosecution and label each case accordingly. When we take this approach, we find that prosecutions involving employees as environmental crime victims can be organized around four primary themes: hazardous substances crimes, air pollution crimes, water pollution crimes, and state environmental crimes. We explore case examples of each category below.

Hazardous Substances Crimes

In 21 prosecutions or about 41% of cases where workers were environmental crime victims, we find the prosecution to revolve around hazardous substances crimes. This category is broad and includes crimes involving pesticides, chemicals, and hazardous waste crimes. The prosecutions focused on violations of CERCLA, FIFRA, HMTA, RCRA, and other crimes. We provide case examples with a discussion of the prosecution of James Blair, John F. Clement, Edgar Wilson, Kelly Steen, and Signal Peak Energy.

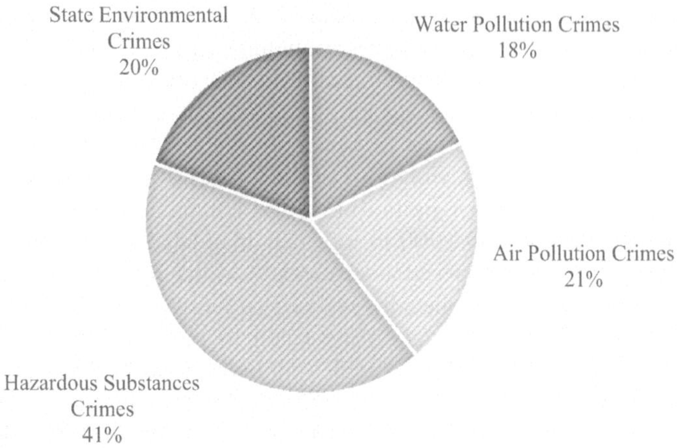

Fig. 3.6 Themes in prosecutions with employees as environmental crime victims (*Source* EPA Summary of Criminal Prosecutions Database)

Prosecutors indicted James Blair after a company employee was injured when the Smith Tank company chose to dump and burn hazardous waste, rather than disposing of it properly. Blair was prosecuted for failure to notify of the release of a hazardous substance under CERCLA and was sentenced to serve 12 months of probation and pay a $10,000 fine.[20] John F. Clement, owner of Casa Famoso Packaging, was prosecuted after ordering the spraying of a restricted use pesticide while employees were working and directing them to continue working, despite a 12-hour reentry period for the application.[21] Workers sought medical attention for headaches, rashes, and sore eyes and throats. Clement was prosecuted for violations of FIFRA (knowingly using a pesticide in a manner inconsistent with its labeling) and was sentenced to pay a $1,000 fine. Prosecutors charged Edgar Wilson and Rockwell International after the company disposed of rocket propellant illegally that resulted in an explosion that killed two employees and injured another.[22] Rockwell pleaded guilty to violations of RCRA and was sentenced to pay over $6.5 million in fines, where Wilson was ordered to serve 12 months of probation and pay a $45,000 fine.

Prosecutors charged Kelly Steen, a driver for Woody's Trucking, for illegal transportation of hazardous waste without a permit.[23] Steen transported flammable natural gas concentrate under the guise of it being a non-hazardous substance, but when offloading the material, it caught fire injuring three employees. The defendant was charged with violations of the HMTA and was sentenced to 36 months of probation and to pay a $2,000 fine. In a related prosecution, Donald E. Wood Jr. and Woody's Trucking were prosecuted, resulting in 12 months and one day in prison for Wood and three years of supervised release, while the company was ordered to serve four years of probation and pay $644,689.70 in restitution for damages to the facility.[24]

Officials charged Signal Peak Energy for a number of offenses related to worker safety and environmental crimes.[25] The company ordered workers to illegally dispose of hazardous waste, paying bribes to a worker not to report an injury, and pressuring a worker who was injured working in the mine to not report the injury as work related after being transported to the hospital.[26] In the only case in our data of an environmental crime occurring against workers in the mining industry, the company was charged with violations of the Mine Safety and Health Act and was sentenced to serve three years of probation, pay a $1 million fine, and $400 in restitution to the injured worker.[27] Dale Musgrave, the vice president of underground operations for Signal Peak, pressured workers not to report workplace accidents and conspired with others to underreport such accidents.[28] He was sentenced to serve three years of probation and pay a $20,000 fine.[29]

Air Pollution Crimes

In 11 prosecutions or 21% of cases where workers were victims of environmental crimes, we find the primary crime at the center of the prosecution focuses on air pollution crimes. These cases were varied and included explosions, releasing pollutants into the ambient air, and illegal asbestos abatement and removal cases. In all of these cases in our data, defendants were prosecuted for violations of the CAA. We provide case examples with the prosecution of Ashland Oil, Chemical & Metal Industries, KMTEX, Mark Hurst, Airosol Company, and Big Lake Gas Plant.

Prosecutors charged Ashland Oil, Minnesota, after an explosion occurred at their Ashland Petroleum Refinery in St. Paul Park, Minnesota.[30] Five employees were injured in the explosion. The company was charged for negligent endangerment under the CAA and making false statements. Ashland was sentenced to serve 60 months of probation, pay a $1.5 million criminal fine, $3.9 million in payments, and $3,750,000 in restitution to the victims and fire department that responded to the explosion. Officials indicated Chemical & Metal Industries for violations of RCRA and the CAA after a mislabeled container released spent antimony pentachloride killing a worker.[31] The company was sentenced to serve 24 months of probation, pay a $1 million fine, and pay $2 million in restitution to the victim's estate.

Prosecutors charged KMTEX for violations of the CAA after a tank explosion killed one worker and severely injured two other workers.[32] The company and its subsidiaries were sentenced to pay $3.5 million in fines and community service payments. Government prosecutors charged Mark Hurst, plant manager of Custom Carbon Processing after he directed employees to accept shipments of natural gas condensate to thin slop oil, despite warnings from the plant's manager.[33] When offloading the gas, it exploded, injuring three workers and causing extensive damage to the plant. Hurst was sentenced to serve two years of probation, fined $5,000, and ordered to pay $12,000 in restitution. CEO Peter Margiotta was sentenced to serve 18 months in prison, three years of supervised release, to pay a $50,000 fine and $100,000 in restitution.[34]

The Airosol Company was charged after a fire and explosion injured a worker at their manufacturing facility in Neodesha, Kansas.[35] The company was sentenced to serve 12 months of probation and to pay a $1,000 fine. Government officials charged Big Lake Gas Plant for releasing 525 pounds of hydrogen sulfide into the ambient air.[36] An employee died as the result of his exposure to the toxic gas and another was injured. The company pleaded guilty to one count of negligent endangerment under the CAA and was sentenced to pay a $3 million criminal fine.[37]

Water Pollution Crimes

In nine prosecutions or about 18% of prosecutions where workers were environmental crime victims, we find the primary crime in the prosecution focuses on water pollution. All but one of these cases involves violations of the CWA and involves illegal discharges of material to the navigable waters of the United States and a number involve a related explosion on an offshore oil platform that also caused an oil spill. Cases include the prosecution of Poling Transportation Company, Halliburton Energy Services, Inc., Wood Group Production Services Network, Black Elk Energy, Grand Isle Shipyard, and Patrick Dooley.

Prosecutors charged Poling Transportation after a fire broke out when material was offloaded from a barge to a tanker truck and two employees caught fire.[38] Poling was charged with violations of the CWA and sentenced to serve 48 months of probation and to pay $126,400 in fines. Officials indicated Halliburton Energy Services for destroying evidence during the investigation of BP, Transocean, and other companies in connection with the Deepwater Horizon disaster that killed 11 workers.[39] In the only case in this category, where defendants were not charged under the CWA, Halliburton pleaded guilty to destroying evidence and was sentenced to pay a $200,000 fine and to serve three years of probation.[40]

Prosecutors indicted Wood Group Production Service Network for its role in an explosion at one of Black Elk Energy's offshore oil platforms that resulted in the death of three workers and numerous injured, as well as an oil spill in the Gulf of Mexico.[41] The company was charged with CWA violations for the illegal discharge of oil, false reporting for showing safety inspections that never occurred, and ordered to pay $9.5 million in monetary penalties. Government officials charged Black Elk Energy for their role in the West Delta 32 explosion and oil spill.[42] The company was ordered to pay $4.2 million in monetary penalties. Prosecutors charged Grand Isle Shipyard for violations of the CWA for its role in the West Delta 32 explosion and oil spill.[43] The company was ordered to pay a $500,000 criminal fine and serve 36 months of probation.[44] Patrick Dooley, owner of AAA Liquidating Services, was charged with CWA violations after he ordered an employee to dump chemicals down the drain, resulting in chlorine gas exposure that caused the employee to be transported to the hospital. Dooley was sentenced to 33 months of incarceration and 36 months of probation.[45]

State-Level Environmental Crimes

In 10 prosecutions or 20% of our cases where workers were victims of environmental crime, we find the primary crime in the prosecution centers on violations of state environmental laws. Eight of the ten prosecutions are related to the Santa Clara Waste Water Company (SCWWC) prosecution.[46] The company and its officials were convicted of a number of offenses after a fire and explosion occurred at the facility. Those unrelated to that prosecution include the prosecution of the Joseph Damato Paper Stock Company and Environmental Enterprises.

Prosecutors charged the Joseph Damato Paper Stock Company after 13,00 aerosol cans were crushed causing an explosion that killed one employee and seriously injured another.[47] The company and its owners were charged with aggravated assault, failure to complete a manifest, and violations of the New Jersey Air Pollution Control Act. The company was sentenced to pay a $250,000 fine and owners and co-defendants Joseph F. Damato and Joseph P. Damato were sentenced to serve 36 months of probation and pay $435 in costs and 36 months of probation and $485 in costs, respectively. Government officials charged Environmental Enterprises after a fire and deflagration occurred at the company's facility in Cincinnati, Ohio. Two of the workers at the facility were burned and one died the next day. The company was charged with negligent homicide and ordered to pay a $5,000 fine.[48]

Punishment and Significant Prosecutions

In the final section of the analysis, we turn to describing punishment and sentencing patterns in environmental crime prosecutions, where employees were victims of environmental crimes. We first outline aggregate sentencing trends, followed by a discussion of significant prosecutions that influence such patterns. We conclude with a discussion of restitution paid to victims in these cases.

In Table 3.1, we illustrate total penalties at sentencing for prosecutions with employees as environmental crime victims. In the top-left quadrant of the table, we show total monetary penalties and assessments ordered at sentencing. We find that individual defendants were collectively sentenced to pay over $4.6 million in monetary penalties at sentencing. Companies were ordered to pay over $535 million in penalties for a grand total exceeding $540 million in penalties at sentencing. In the upper-left quadrant, we show total years of probation assessed in these prosecutions. Individual defendants were collectively assessed 102 years of probation and companies were assessed 67 years of probation at sentencing for a cumulative total of 169 years of probation. In the lower-right quadrant, we show total incarceration time assessed to defendants at sentencing, which totaled 57 years in the analysis. Defendants were also sentenced to serve 210 hours of community service at sentencing.

Table 3.1 Total penalties assessed at sentencing for prosecutions with employees as environmental crime victims

Monetary penalties Individuals: >$4.6 Million Companies: >$535 Million **Years of incarceration** 57	**Years of probation** Individuals: 102 Companies: 67 **Hours of community service** 210

*Numbers are rounded
Source EPA Summary of Criminal Prosecutions Database

In Table 3.2, we explore cases of significant probation assessed at sentencing for prosecutions with employees as victims of environmental crime. Three cases make up 468 months or 39 years of probation assessed at sentencing. These cases make up roughly 23% of total probation in our data. Nebraska Railcar Cleaning Services was prosecuted, along with president Steven Michael Braithwaite and Adam Thomas Braithwaite, when flammable cases inside a rail tanker car ignited killing two workers and injuring another.[49] The company had been cited by OSHA for safety violations for confined spaces at work, such as railcars. Steven Braithwaite lied to investigators that the company had been testing for benzene and submitted falsified documents to OSHA. Charged with violations of RCRA and other crimes, Steven Braithwaite was sentenced to serve 30 months in prison and pay $100,000 in restitution. Adam Braithwaite was sentenced to serve a year and a day in prison and pay $100,000 in restitution. Both of the defendants and the company were ordered to serve five years of probation and the company was ordered to pay a $21,000 fine.[50]

Table 3.2 Cases of significant probation assessed at sentencing for prosecutions with employees as environmental crime victims

Defendant	*Fiscal year*	*Crime*	*Total probation (years)*
Nebraska Railcar Cleaning Services	2022	Hazardous substances	15
Texas Oil & Gathering, Inc.	2010	Hazardous substances	13
Mark Avila	2020	State law	11

Source EPA Summary of Criminal Prosecutions Database

Texas Oil & Gathering was sentenced to pay a $80,000 fine and serve 36 months of probation for illegal transport of hazardous waste and for conspiracy.[51] John Kessel, owner of the company, was sentenced to pay a fine of $50,000 and serve five years of probation for conspiracy and violations of the SDWA. Edgar Pettijohn, the company manager, was sentenced to serve five years of probation. The company, located in Rosharon, Texas, operation injection wells to dispose of oil and natural gas waste. In January, 2003, a vapor cloud caused an explosion during the offloading process of the waste, killing two truck drivers and injuring three other employees.[52]

Mark Avila was prosecuted in conjunction with the broader prosecution of the SCWWC.[53] One of the more significant state environmental crimes in our analysis, a vacuum cargo tank trailer containing hazardous waste exploded at the company's facility in Santa Paula, California, causing a chemical spill that further ignited a number of hazardous materials at the facility. A mandatory evacuation was ordered in a three-mile radius of the facility; dozens of bystanders, along with firefighters and company employees, were injured. The company had been engaging in the illegal storage and disposal of hazardous waste. Facility manager Mark Avila and co-defendant Brock Baker were sentenced to serve 36 months of probation, two months of suspended jail time, and pay a $150 fee. Human resources manager, Marlene Faltemier, was sentenced to pay over $2.6 million in restitution to victims and serve five years of probation.[54]

In Table 3.3, we describe cases of significant incarceration assessed at sentencing for cases where employees were environmental crime victims. These cases represent 41 years of incarceration in our data on employees as victims, or about 72% of total incarceration in this case. Like probation, incarceration is greatly affected in this case by a few significant prosecutions. We discuss the prosecution of Alan Elias, Atlantic States Cast Iron Pipe, and Lloyd Robl for context.

Alan Elias, owner of Evergreen Resources in southeastern, Idaho, directed employees to enter a 25,000-gallon tank to remove sludge that contained toxic cyanide.[55] Not provided proper instruction or safety equipment, one employee was overcome by the fumes and suffered irreversible brain damage as a result of Elias's actions.[56] Elias continued to direct the employees to bury some 8,000 gallons of toxic sludge. For giving false statements and violations of RCRA, Elias was charged with knowing endangerment of his employees and was sentenced to serve 204 months of incarceration, 36 months of probation, to pay $364,750 in

Table 3.3 Cases of significant incarceration assessed at sentencing for prosecutions with employees as environmental crime victims

Defendant	*Fiscal year*	*Crime*	*Total incarceration (years)*
Alan Elias	2003	Hazardous substances	17
Atlantic States Cast Iron Pipe	2009	Water pollution	12
Lloyd Robl	2019	Air pollution	12

*Numbers are rounded
Source EPA Summary of Criminal Prosecutions Database

restitution to the U.S. EPA and $5.9 million in restitution to his victim.[57] The Elias' prosecution stands as one of the most punitive in our analysis and in the history of environmental crime prosecutions. A stark example of a supervisor directly injuring their employee, the case received national news coverage and was the basis for the book, *The Cyanide Canary*.[58]

Atlantic States Cast Iron Pipe was prosecuted for violations of the CAA and CWA, as well as numerous OSHA safety regulations after multiple workers were injured and one killed at their facility in Phillipsburg, New Jersey.[59] On January 1, 2006, a 34-count indictment named the company and managers for conspiring to violate clean air and water laws and skirt workplace safety laws, as well as impeding federal investigations into the company. The company was ordered to pay a $8 million federal fine, a $12,000 special assessment, and serve 48 months of probation. Four co-defendants at the company were collectively sentenced to serve 147 months in prison, as well as $2,325 in special assessments.[60]

Lloyd A. Robl, an asbestos abatement contractor in Minnesota and Wisconsin, was prosecuted for fraudulently claiming to be a licensed contractor capable of providing asbestos abatement services.[61] Additionally, Robl falsified a number of documents related to his license, insurance documents, asbestos waste manifests, and air sampling data while conducting illegal abatement jobs. The defendant disposed of the asbestos waste in burn piles and nine victims testified at sentencing, including four methamphetamine addicts that claimed Robl gave them methamphetamines as pay and exposed them to asbestos containing materials on the job with no training or protection. Firefighters were also exposed to asbestos when extinguishing the asbestos-laden burn pits that raged out of control on the defendant's property. Robl pleaded guilty to wire fraud and violations of the CAA and was sentenced to serve 12 years in prison and three years of supervised release.[62]

In Table 3.4, we describe significant monetary penalties assessed at sentencing in prosecutions, where employees were victims of environmental crimes. The prosecution of three corporations, Transocean, BP Products North America, and Honeywell International, resulted in $462 million in monetary penalties in our data. These three cases make up 86% of all monetary penalties in our analysis of employees as crime victims, repeating a similar pattern found in probation and incarceration trends in the previous two tables.

Transocean was prosecuted for its role in the BP Deepwater Horizon disaster in the Gulf of Mexico. Members of the company's crew on the Deepwater Horizon oil rig were negligent in failing to predict and control the blowout in the Macondo well that resulted in the death of 11 rig workers and the largest maritime oil spill in U.S. history.[63] The company pleaded guilty to violating the CWA and was ordered to serve five years of probation and to pay $400 million in criminal penalties.[64] The company also agreed to pay $1 billion in civil penalties.[65]

BP Products North America was prosecuted for violations of the CAA after an explosion at the company's Texas City Refinery killed 15 workers and injured some 170 others.[66] The company was sentenced to pay a $50 million federal fine and serve 36 months of probation.[67] The case became a central part in the PBS documentary, *The Spill*, which cataloged the history of BP's management systematically underinvesting in proper safety measures.[68] Other cases included in the EPA database that preceded the Texas City case and the problems leading to it include the prosecution of BP Exploration Alaska (BPXA) in 2000 and 2008, for violations of the

Table 3.4 Cases of significant monetary penalties assessed at sentencing for prosecutions with employees as environmental crime victims

Defendant	*Fiscal year*	*Crime*	*Total monetary penalties*
Transocean	2014	Water pollution	$400 Million
BP Products North America	2009	Air pollution	$50 Million
Honeywell International	2007	Air pollution	$12 Million

*Numbers are rounded

Source EPA Summary of Criminal Prosecutions Database

CWA and RCRA.[69] The previously discussed prosecution of Honeywell International resulted in a $12 million in penalties.[70] In another significant case, prosecutors charged the Motiva Corporation for violations of the CAA and CWA after a storage tank ruptured, causing the leakage of 300,000 gallons of spent sulfuric acid on the ground that caught fire and resulted in injuries to eight workers. The company was ordered to pay a $10 million fine and a $525 special assessment.[71]

In the final section of the analysis in the chapter, in Table 3.5, we explore the percentage of cases where restitution was awarded to victims in environmental crime prosecutions where employees were victims of environmental crimes. We organize restitution cases by number of prosecution and percentage of prosecutions, across the major themes we uncovered in prosecutions in Fig. 3.6. Overall, we find a total of 17 prosecutions or about 35% of prosecutions involving employees as victims, note that restitution was assessed at sentencing.

In the upper-left quadrant, we show that six prosecutions or about 13% of cases of hazardous substances crimes involved restitution to victims. In the upper-right quadrant, seven prosecutions or about 15% of cases involved restitution. In the lower-left quadrant, four prosecutions or roughly 8% of prosecutions involving air pollution crimes also assigned restitution at sentencing. We found no cases of water pollution crimes that assessed restitution to victims at sentencing.

It is difficult to assess the full extent of restitution ordered by the courts to employees that were victims of environmental crime. The case summaries can be unclear about whether restitution was to the victims, to other companies, government, and other injured parties. Much of the restitution would come from the SCWWC prosecution and the full extent of that restitution for all defendants is unclear in all of the cases

Table 3.5 Percentage of cases where restitution was awarded to victims for prosecutions with employees as environmental crime victims

Hazardous substances crimes 6 Prosecution (13% of cases)	**State environmental crimes** 7 Prosecutions (15% of cases)
Air pollution crimes 4 Prosecutions (8% of cases)	**Water pollution crimes** 0 Prosecutions (0% of cases)

Source EPA Summary of Criminal Prosecutions Database

in our data. Moreover, while the courts may have imposed restitution, whether victims see some or all of it is difficult to know and restitution may be thrown out on appeal, which was the case with the prosecution of Alan Elias, where the court awarded \$6 million in restitution to victims, making this the largest judgment to victims in our data. Our best estimate of total restitution strictly from the EPA data is about \$24.9 million including the Elias verdict and \$18.9 million if we excluded it from the total, with over \$8.6 million ascribed to the SCWWC, making it responsible for about 46% of restitution in our analysis of employees as environmental crime victims.

Conclusion

Examining patterns and themes in prosecutions involving employees as environmental crime victims has yielded a few important empirical and conceptual insights. First, prosecutions of environmental crimes in the workplace are rate in our data. We uncover only 51 prosecutions that in our judgment focus on employees as victims of environmental crimes. These prosecutions increased over the last five years in our analysis, whereas they were quite sporadic prior to the 2000s. They were also geographically centered, with over 40% of prosecutions taking place within California, Louisiana, and Texas.

A second implication of our findings is that a number of the prosecutions under analysis involve criminal behavior and have serious consequences. Roughly half of the prosecutions analyzed in this chapter involve at least one Title 18 crime, such as false statements conspiracy, and fraud. Employees also suffered grave consequences from these criminal acts. Forty percent of prosecutions were of cases where employees were injured. Death occurred in about 16% of prosecutions and death and injury occurred in 27% of prosecutions. Evacuation and injury took place in about 14% of prosecutions.

A third finding of importance is the significant amount of monetary penalties and other punishments assessed at sentencing, as well as restitution paid to victims. Over \$540 million in monetary penalties were assessed to all defendants at sentencing, along with 169 years of probation and 57 years of incarceration. While impressive, it is important to note these trends are significantly affected by a few significant prosecutions. Three prosecutions are responsible for 24% of probation totals. Three different cases are responsible for over 70% of total incarceration in our analysis. Finally, the prosecution of three large corporations resulted in the vast majority of all monetary penalties in the analysis. Victim restitution is much harder to estimate with our data, but we find that about a third of the case summaries mention restitution imposed at sentencing. Of those cases, the prosecution of officials at the SCWWC makes up about 46% of restitution that totals approximately $18.9 million.

Environmental crimes in the workplace can come with stiff penalties in some cases, but those significant penalty cases draw out most of the punishments in our data. While the existence of these cases shows the fortitude of EPA-CID criminal investigators and DOJ prosecutors to pursue difficult cases and obtain large outcomes, these cases are few. In most of these prosecutions, we would argue there is a good chance the penalties and punishments are robust in terms of specific deterrence effects to the individuals and companies (they can be harsh and we do not see many repeat offenders), but whether they pose broader deterrent effects for related industries or environmental crime in general is hard to know.

Notes and References

1. U.S. Department of Justice, 2013, Former President of Port Arthur, Texas Chemical Company Sentenced for Federal Crimes Related to Employee Deaths, Available from: https://www.justice.gov/opa/pr/former-president-port-arthur-texas-chemical-company-sentenced-federal-crimes-related-employee.
2. Sandy Smith, Former President of Chemical Company Sentenced for Federal Crimes Related to Employee Deaths, Available from: https://www.ehstoday.com/standards/osha/article/21916272/former-president-of-chemical-company-sentenced-for-federal-crimes-related-to-employee-deaths.
3. U.S. Occupational Safety and Health Act (29 U.S.C., 651, 1970).
4. Dan Wallach, 2013, Business Owner Sent to Prison in Death of Man, Available from: https://www.chron.com/news/houston-texas/houston/article/Business-owner-sent-to-prison-in-death-of-man-4933934.php.
5. Matthew Lawrence Bowman (E.D. Texas 1:12-CR-42, 2014).
6. In nine prosecutions, we find that both employees are victims of environmental crime and the public is also injured. We felt the cases in this chapter hinged on the employee victimization

and include these cases here too, even though injuring the public is a secondary consequence of the crime and often not nearly as severe. Also, they involve only one distinct and eight related prosecutions. In the prosecution of the Airosol Company, evacuations occurred in the surrounding community: Airosol Company (D. Kansas 6:21CR10074, 2022).

In eight cases, related to the Santa Clara Waste Water Company (SCWWC), evacuation of the facility occurred as the second effect of the crime as did injuries to first responders. Charles Ray Mundy (California 2015023881, 2018); Dean Michael Poe (California 2015023881, 2018); David Joseph Wirsing (California 2015023881, 2018); Douglas Brian Edwards (California 2015023881, 2017); William James Mitzel (California 2015023881, 2017); Kenneth Douglas Griffin (California 2015023881, 2017); Green Compass Environmental (California 2015023881, 2017); Mark Avila (California 2015023881, 2020).

7. Albert David Hajduk (D. Colorado 04-CR-00346-LTB, 2006). A worker and the Persigo POTW was injured in the case and the company was ordered at sentencing to pay $350,000 to the injured worker, along with a $40,000 fine, and a $125 special assessment.
8. U.S. EPA, 1999, Chemical Company Pleads Guilty to Hazardous Waste Crime That Killed Ohio Employee, Available from: https://www.epa.gov/archive/epapages/newsroom_archive/newsreleases/6afcd1da58baa75685256803006eed21.html.
9. Lancaster Synthesis, Inc. (S.D. Ohio CR-1-99-0085, 2000).
10. Honeywell International (M.D. Louisiana 07-31-FJP-SCR, 2007).
11. Reuters, 2007, Honeywell Sentenced for Fatal Toxic La. Incident, Available from: https://www.reuters.com/article/us-honeywell-sentenced-toxic-idUSWEN0972200709 13.
12. Durbin Hartell (D. Oregon 08-266-KI, 2009).
13. U.S. Department of Justice, 2008, Oregon Corporation and Its President Sentenced for Hazardous Waste Violations, Available from: https://www.justice.gov/archive/opa/pr/2008/June/08-enrd-506.html.
14. Spencer Environmental, Inc. (D. Oregon 07CR381-02-HA, 2008).
15. U.S. Department of Justice, 2013, Owner of Colorado Aircraft Painting Company Pleads Guilty to Unlawfully Treating

Hazardous Waste, Available from: https://www.justice.gov/opa/pr/owner-colorado-aircraft-painting-companypleads-guilty-unlawfully-treating-hazardous-waste.

16. Norman Teltow (D. Colorado 13-CR-00027-MSK, 2014).
17. EPA and the DOJ also reached a civil settlement with the company:

    U.S. EPA, 2024, MFA Incorporated and MFA Enterprises Incorporated, Settlement Information Sheet, Available from: https://www.epa.gov/enforcement/mfa-incorporated-and-mfa-enterprises-incorporated-settlement-information-sheet.
18. MFA Incorporated (E.D. Missouri 407CR00550TIA, 2008).
19. Miller Industries, Inc. (D. Maine 2:07-CR-12, 2007).
20. James Blair (E.D. Texas 6-94 CR, 1995).
21. John F. Clement (E.D. California MFG-98-2119, 1998).
22. Edgar Wilson (C.D. California CR 99-423(A)-ALL, 2004).
23. Kelly Steen (D. Montana CR 14-111-BLG-SPW, 2015).
24. Donald E. Wood Jr. (D. Montana CR 17-138-BLG-SPW, 2019).

    United States Attorney's Office, 2018, Baker Man and Trucking Company Ordered to Pay $1.29 Million, Face Prison for Fraud and Violating Laws Regulating Transporting Hazardous Materials, Available from: https://www.justice.gov/usao-mt/pr/baker-man-and-trucking-company-ordered-pay-129-million-face-prison-fraud-and-violating.
25. Signal Peak Energy, LLC. (D. Montana 1:19-CR-00159, 2022).
26. The company has become part of a broader investigation into the malfeasance of its owners and managers,

    Hiroki Tabuchi, 2023, A Faked Kidnapping and Cocaine: A Montana Mine's Descent into Chaos, New York Times, Available from: https://www.nytimes.com/2023/01/13/climate/signal-peak-mine-coal.html.
27. Federal Mine Safety and Health Act (P.L. 95-164, 1977).

    United States Attorney's Office, District of Montana, 2022, Signal Peak Energy Fined $1 Million, Sentenced to Three Years of Probation for Violating Environmental Worker Safety Regulations at Roundup Coal Mine, Available from: https://www.justice.gov/usao-mt/pr/signal-peak-energy-fined-1-million-sentenced-three-years-probation-violating.
28. Dale Musgrave (D. Montana 1:19-cr-00159-SPW, 2022).

    United States Attorney's Office, 2021, Signal Peak Energy Admits Violating Health, Safety Regulations at Roundup Area Coal Mine, Available from: https://www.justice.gov/usao-mt/pr/signal-peak-energy-admits-violating-health-safety-regulations-roundup-area-coal-mine.

29. Other prosecutions in this category include: OEA, Inc. (D. Colorado 94 CR 75B, 1995). A manufacturer of automotive airbags, the company illegally transported hazardous materials off-site through open burning that caused severe burns to five individuals. Carlton Finley, project manager for Donaldson Enterprises, (D. Hawaii CR17-00178 SOM, 2017) was prosecuted after a fire occurred at the Waikele self-storage facility in Waipahu, Hawaii, that contained fireworks. Five company personnel died as the result of the accident. C&J Well Services (D. North Dakota 1:19-CR-00079-DLH, 2019) was prosecuted after a worker was directed to weld an unclean tank that exploded and killed him. The company was ordered to pay a $500,000 fine and $1.6 million in restitution to the victim. Stephan Reisinger (D. North Dakota 1:19-cr-240, 2022), the maintenance manager at the plant, was also prosecuted for his role in the worker's death. Loren Jacobson (D. Idaho 4:21-cr-00149, 2022) was prosecuted for an explosion that occurred during a cargo tanker repair that severely injured one of his employees. Jacobson owned KCCS, Inc., a tanker testing and repair company. John Myre (D. Idaho 3:14-CR-27-EJL, 2014) was prosecuted for negligent endangerment under the CAA. Myre was a supervisor for the Dakota Southern Railroad and ordered employees to use cutting torches on beams that were coated in lead-based paint. One of the workers was hospitalized for acute lead poisoning and the others had high levels of lead in their blood. The defendant was ordered to pay a $3,000 criminal fine, serve 90 hours of community service, and serve three years of supervised release.
30. Ashland Oil (W.D. Kentucky 88-146, 2003).
31. Chemical & Metal Industries (M.D. Louisiana08-28-JUP-CN/08-28-RET-CN, 2011).
32. KMTEX (E.D. Texas 1:16-CR-00075-001, 2017).
33. Mark Hurst (D. Montana CR 17-143-BLG-SPW, 2020).

    United States Attorney's Office, 2019, Ex-Manager of Wibaux Oil Processing Plant that Exploded Sentenced for Clean Air Act Violations, Available from: https://www.justice.gov/usao-mt/pr/ex-manager-wibaux-oil-processing-plant-exploded-sentenced-clean-air-act-violation.
34. Peter Margiotta (D. Montana CR 17-143-BLG-SPW, 2020).

United States Attorney's Office, 2020, Former Custom Carbon Processing Official Sentenced to Prison, Fined for Violating Clean Air Act for Actions Leading to Explosion of Wibaux Oil Processing Plant, Available from: https://www.justice.gov/usao-mt/pr/former-custom-carbon-processing-official-sentenced-prison-fined-violating-clean-air-act.

35. Airosol Company (D. Kansas 6:21CR10074, 2022).
36. Big Lake Gas Plant, LP (N.D. Texas 6:21-CR-00057-H-BU, 2022).
37. Matt Trammell, 2022, Big Lake Gas Plant Forced to Pay Millions After Being Found Criminal Negligent, Available from: https://sanangelolive.com/news/business/2022-02-18/big-lake-gas-plant-forced-pay-millions-after-being-found-criminal-negligent.
38. Poling Transportation Corp. (E.D. New York, 1997).
39. Halliburton Energy Services (E.D. Louisiana 2:13-CR-00165-JTM-KWR, 2013).
40. U.S. Department of Justice 2017, Company to Pay $9.5 Million for False Reporting of Safety Inspections and Clean Water Act Violations That Led to Explosion in Gulf of Mexico, Available from: https://www.justice.gov/opa/pr/company-pay-95-million-false-reporting-safety-inspections-and-clean-water-act-violations-led.
41. Wood Group Production Services Network (E.D. Louisiana 15-197 H, 2017).
42. Black Elk Energy (E.D. Louisiana15-197 SECT H, 2017).

    U.S. Department of Justice, 2017, Black Elk Energy Offshore Operations LLC. Convicted of Worker Safety and Clean Water Act Violations in Connection to Offshore Explosion, Available from: https://www.justice.gov/opa/pr/black-elk-energy-offshore-operations-llc-convicted-worker-safety-and-clean-water-act.
43. Grand Isle Shipyard (E.D. Louisiana 15-197 SECT H, 2019).
44. U.S. Department of Justice, 2015, Three Companies and Three Individuals Charged in Fatal 2012 Gulf of Mexico Oil Drilling Platform Explosion, Available from: https://www.justice.gov/opa/pr/three-companies-and-three-individuals-charged-fatal-2012-gulf-mexico-oil-drilling-platform.
45. Patrick Dooley (W.D. Washington CR11-252MJP, 2012).
46. The following prosecutions took place as the result of the SCWWC case in our data:

    Charles Ray Mundy (California 2015023881, 2018); Dean Michael Poe (California 2015023881, 2018); David Joseph Wirsing (California 2015023881, 2018); Douglas Brian Edwards

(California 2015023881, 2017); William James Mitzel (California 2015023881, 2017); Kenneth Douglas Griffin (California 2015023881, 2017); Green Compass Environmental (California 2015023881, 2017); Mark Avila (California 2015023881, 2020).

47. Joseph Damato Paper Stock Company (D. New Jersey 02 04 0446 I, 2003).
48. Environmental Enterprises (Ohio B1406608A, 2017).
49. U.S. Department of Justice, 2022, Omaha Railcar Cleaning Company and Its Owners Sentenced for Violating Environmental and Worker Safety Laws Resulting in Workers' 2015 Deaths, Available from: https://www.justice.gov/opa/pr/omaha-railcar-cleaning-company-and-its-owners-sentenced-violating-environmental-and-worker.
50. Adam Thomas Braithwaite (D. Nebraska 8:18CR216-002, 2022).
51. Texas Oil & Gathering, Inc. (S.D. Texas 4:07CR00466-001, 2010).
52. U.S. Department of Justice, 2009, Oil Company and Two Executives Plead Guilty to Environmental Crimes, Available from: https://www.justice.gov/opa/pr/oil-company-and-two-executives-plead-guilty-environmental-crimes.
53. Mark Avila (California 2015023881, 2020).
54. SCVNEWS, 2019, Former Wastewater Officials Sentenced on Charges Related to Hazardous Materials Scheme, Available from: https://scvnews.com/former-wastewater-officials-sentenced-on-charges-related-to-hazardous-materials-scheme/.
55. AP, 2000, Businessman Sentenced to Prison for Exposing Worker to Cyanide, Available from: https://www.nytimes.com/2000/04/30/us/businessman-sentenced-to-prison-for-exposing-worker-to-cyanide.html.
56. U.S. EPA, 2000, Idaho Man Receives 17-Year Prison Sentence for Severe Environmental Crimes, Available from: https://www.epa.gov/archive/epapages/newsroom_archive/newsreleases/60100e7844b12528852568d600583d2b.html.
57. Alan Elias (D. Idaho CR 98-070-BLW, 2003). The circuit court vacated the restitution imposed by the lower court for the victim in the case.
58. Kurt Friedemann, 2000, Boss Must Pay for Poisoning Employee, Available from: https://www.hcn.org/issues/176/5696.

Robert Dugoni and Joseph Hilldorfer, 2014, The Cyanide Canary: A True Story of Injustice, Brown Books.

59. Atlantic States Cast Iron Pipe (D. New Jersey 3:03cr852, 2009).
60. U.S. Department of Justice, 2009, Cast Iron Pipe Manufacturer Sentenced for Environmental Crimes and Worker Safety Violations, Available from: https://www.justice.gov/opa/pr/cast-iron-pipe-manufacturer-sentenced-environmental-crimes-and-worker-safety-violations.
61. Lloyd A. Robl (W.D. Wisconsin 3:18-CR-00136-WMC, 2019).
62. U.S. Department of Justice, 2019, Former Asbestos Abatement Contractor Sentenced for 12 Years for Violating the Clean Air Act, Available from: https://www.justice.gov/usao-wdwi/pr/former-asbestos-abatement-contractor-sentenced-12-years-violating-clean-air-act.
63. U.S. Department of Justice, 2013, Transocean Agrees to Plead Guilty to Environmental Crime and Enter Civil Settlement to Resolve U.S. Clean Water Act Penalty Claims from Deepwater Horizon Incident. Available from: https://www.justice.gov/opa/pr/transocean-agrees-plead-guilty-environmental-crime-and-enter-civil-settlement-resolve-us.
64. Transocean LTD (E.D. Louisiana 2:13-CR-00001-JTM-SS, 2014).
65. U.S. EPA, 2024, Transocean Settlement, Available from: https://www.epa.gov/enforcement/transocean-settlement.
66. Other sources give a figure of 180 injuries.

    U.S. Chemical Safety Board, 2023, BP America Refinery Explosion, Available from: https://www.csb.gov/bp-america-refinery-explosion/.
67. BP Products North America (S.D. Texas 4:07-CR-434, 2009).
68. PBS Frontline, The Spill, 2010, Available from: https://www.pbs.org/wgbh/pages/frontline/the-spill/.
69. BP Exploration-Alaska (BPXA (D. Alaska A99-0141, 2000), also see Doyon Drilling, Inc. (D. Alaska A98-082-JKS, 2000); BP Exploration Alaska, Inc. (D. Alaska 3:07-CR-00125-TMB, 2008).
70. (Honeywell International (M.D. Louisiana 07-31-FJP-SCR, 2007).
71. Motiva Corporation (D. Delaware 05-CR-021-SLR, 2005).

CHAPTER 4

When Companies Victimize the Public

Abstract In this chapter, we explore the way that companies victimize the public, through our analysis of environmental crime prosecutions within the United States. From 2,828 total prosecutions, we find 36 prosecutions where a company victimized the public either in the form of causing an evacuation, injury, death, or a combination of these actions. In the first section of the chapter, we mirror the flow from the previous chapter, where we explore patterns in prosecutions over time that involve companies victimizing the public. We organize prosecutions across broader themes in the second section and illustrate a number of case examples for each theme we find in the data. The last section describes trends in sentencing and punishment over time.

Keywords Corporate crime · Public victimized

On July 15, 2009, 7,000 pounds of ammonia gas was offloaded to a Werner Transportation truck at a distribution facility.[1] The hose ruptured, causing 1,800 gallons to spill and a toxic cloud engulfed a passing car

J. Ozymy and M. Jarrell Ozymy, *The U.S. Administrative State and the Protection of Environmental Crime Victims*, Palgrave Studies in Green Criminology, https://doi.org/10.1007/978-3-031-62680-7_4

that killed a woman on her way to work and some seven other individuals were hospitalized. Werner Transportation was charged with negligent endangerment under the CAA and was sentenced on June 19, 2013, to pay a $100,000 fine.[2]

Prosecuting companies like Werner Transportation occurs when their negligent or willful actions result in harm to the general public. We find few examples of prosecutions in our analysis of EPA-CID criminal investigations where a company was prosecuted for an environmental crime that resulted in harm to the public. In this chapter, we describe 36 prosecutions in our data, where actions by companies victimized the general public.

Patterns in Prosecutions

We begin the analysis of companies victimizing the public in Fig. 4.1, where we show the number of cases with an identifiable victim, 1983–2022. We show the first prosecutions with a victim occurring in 1989 (1) and 1990 (1). A total of eight cases emerge from the data in the 1990s, eight prosecutions from 2000 to 2009, and 19 prosecutions from 2010 to 2022. A grand total of 36 prosecutions were adjudicated in our analysis, averaging about 0.9 prosecutions per year.

In Fig. 4.2, we describe the number of prosecutions where companies victimized the public by U.S. state and territory, 1983–2022. We find at least one prosecution occurring within 25 U.S. states and territories, with the largest number adjudicated in California (4), Louisiana (3),

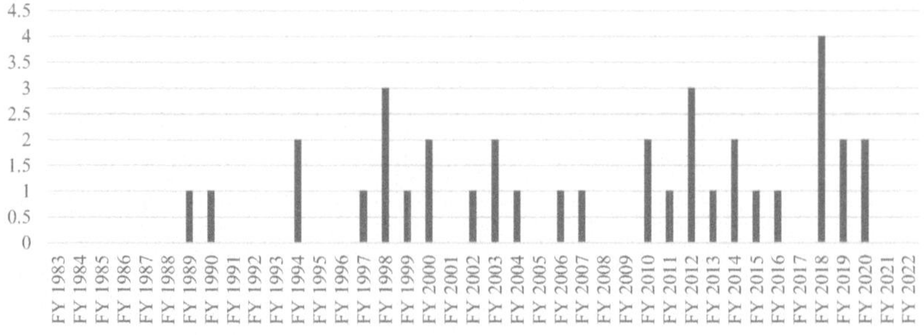

Fig. 4.1 Number of prosecutions where companies victimized the public by EPA fiscal year, 1983–2022 (*Source* EPA Summary of Criminal Prosecutions Database)

and New York (3). Two prosecutions were adjudicated in Florida, Kansas, Washington, and the Virgin Islands. The remaining states, including the U.S. Territory of Puerto Rico, had one case adjudicated in our data. Prosecutions averaged about 1.4 across these states in our analysis.

We describe charging patterns in prosecutions where companies victimized the public in Fig. 4.3. The most common type of crime we encounter in the data is pesticide crimes where at least one defendant was charged with violating FIFRA. These crimes occurred in 11 cases. The second most common crime was a violation of water pollution laws under the CWA, which occurred in eight prosecutions. In eight prosecutions, at least one defendant was charged with hazardous waste crimes in violation of RCRA. In five prosecutions, at least one defendant was charged with air pollution crimes under the CAA. In three prosecutions, at least one defendant was charged with violations of state environmental laws. Finally, in one prosecution at least one defendant was charged with a chemical substances crime under TSCA and in one prosecution, at least one defendant was charged with failing to notify officials of the release of a hazardous substances under CERCLA.

In Fig. 4.4, we describe patterns in criminal behaviors where companies victimized the public. We explore all Title 18 violations and other criminal actions noted in the case summaries. In three prosecutions, at least one defendant was charged with giving false statements on official documents or to government officials. In two prosecutions, defendants were charged with conspiracy. In two prosecutions, defendants were charged

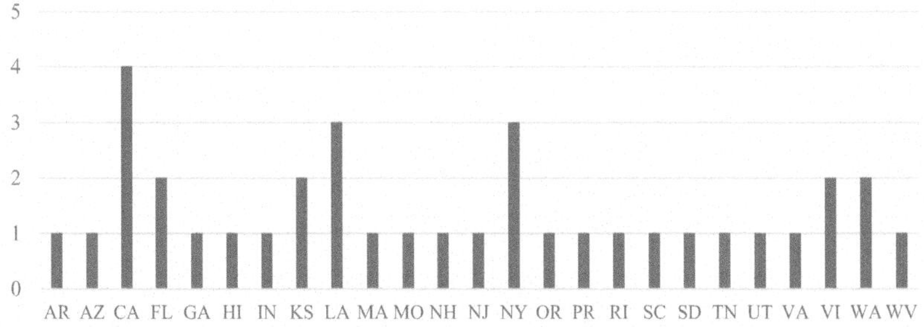

Fig. 4.2 Number of prosecutions where companies victimized the public by U.S. State and Territory, 1983–2022 (*Source* EPA Summary of Criminal Prosecutions Database)

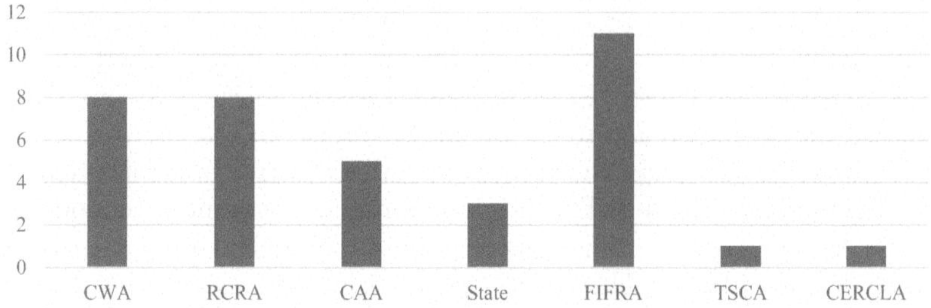

Fig. 4.3 Charging patterns in prosecutions where companies victimized the public, 1983–2022 (*Source* EPA Summary of Criminal Prosecutions Database)

with obstruction. These seven cases account for 16% of cases in our analysis. We provide examples with the prosecution of A-1 Sanitation, CH20, and Harrison Research Laboratory.

Prosecutors charged A-1 Sanitation for dumping toxic solvents into the Wareham, Massachusetts, sewage treatment plant.[3] Chemists inside the plant suffered burning eyes, throat, and dizziness when the hazardous materials made their way into the facility. The company was charged with violations of the CWA for the illegal discharge and making false statements. The company was ordered to serve 36 months of probation and pay a $30,000 fine. CH20, Inc., located near Olympia, Washington, was charged for illegally disposing of chemicals into the city sewer system. The company and four co-defendants were charged with violations of the CWA for the illegal discharges and for conspiracy. Prosecutors charged

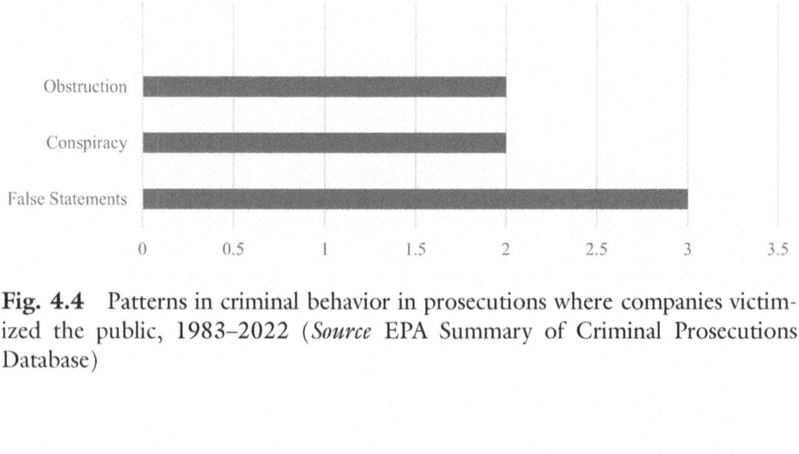

Fig. 4.4 Patterns in criminal behavior in prosecutions where companies victimized the public, 1983–2022 (*Source* EPA Summary of Criminal Prosecutions Database)

Harrison Research Laboratory with for illegally testing the pesticide DEET on over 300 people without explaining the risks.[4]

Figure 4.5 describes the type of victimization experienced by the public, by organized cases across outcomes. In our data, the most common outcome in these prosecutions is injury, where in 16 prosecutions or about 44% of cases analyzed in this chapter, at least one person was injured as the result of a company's environmental crime. In 15 prosecutions, or roughly 42% of prosecutions, at least one individual was evacuated, injured, or a combination of the two in our analysis. In four prosecutions, or about 11% of cases, death resulted from an environmental crime. In one prosecution, or about 3% of cases, death and injury resulted from a company's environmental crime. We provide case examples with the prosecution of Friendly Systems, Inc., JTA Real Estate, and Trinity Railcar Repair.

Prosecutors charged Friendly Systems, Inc., for violations of FIFRA after they illegally used pesticides to sanitize children's toothbrushes for the Rosebud Indian Tribe's Head Start Program and 100 children received blisters and burns in their mouths.[5] The company was sentenced to serve 60 months of probation, pay a $150,000 fine, and pay $10,657 in restitution. Officials indicated JTA Real Estate for obstruction and

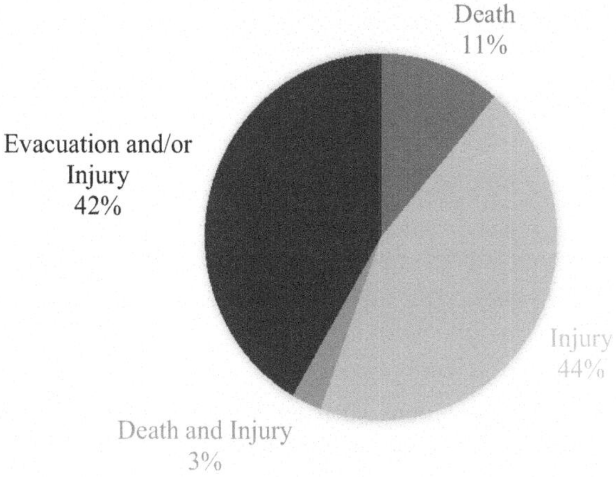

Fig. 4.5 Categories of victimization in prosecutions where companies victimized the public (*Source* EPA Summary of Criminal Prosecutions Database)

violations of TSCA after they failed to provide proper lead-based paint disclosures in one of their units, and a young child died as the result of ingesting lead-based paint.[6] The company was sentenced to pay a $40,000 fine and to serve 36 months of probation.[7] Officials charged Trinity Railcar Repair for violations of the CAA after an ammonia leak in Fort Edward, New York, resulted in the evacuation of 400 residents.[8]

Themes in Prosecutions

In the second section of the chapter, we organize prosecutions where companies victimized the public into general themes. We broadly explore each prosecution and use our best judgment to discern the primary crime that was at the center of the prosecution. Once we categorize prosecutions in this manner, we describe four basic types of crimes that emerge from the data. These include hazardous substances crimes, water pollution crimes, air pollution crimes, and state environmental crimes. We show these percentage of cases that fall within each category in Fig. 4.6 and explore case examples of each below.

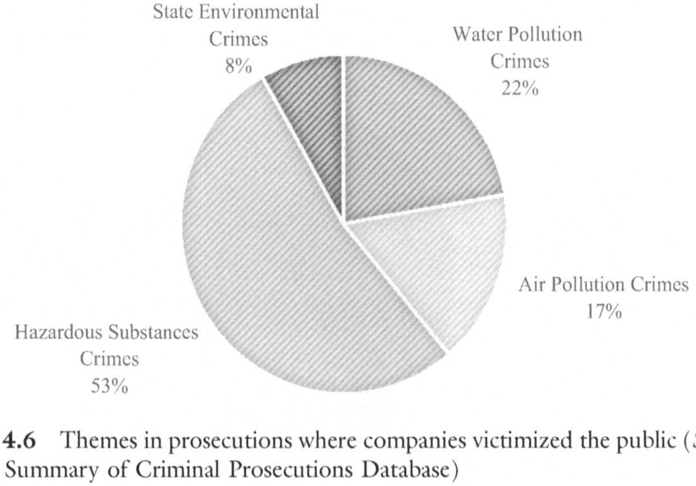

Fig. 4.6 Themes in prosecutions where companies victimized the public (*Source* EPA Summary of Criminal Prosecutions Database)

Hazardous Substances Crimes

In 19 prosecutions or 53% of total cases under consideration in this chapter, the primary crime at the heart of the prosecution centered on a hazardous substance crime. We labeled these cases in this category if they involved illegal behavior regarding pesticides that violated FIFRA and hazardous waste crimes that violated RCRA. We provide case examples with the prosecution of Orkin Exterminating Company, Chemical Commodities, Southern Union Company, DLM Farms, Grenale Leonard Williams, Terminix, USVI, Curtis Technology, Wade Jones Company, Omni Applicators, VC Tank Lines, Mount Whitney Pest Control, Bugman Pest and Lawn, and James Richard Holt.

Prosecutors charged the Orkin Exterminating Company for misuse of Vikane in a home located in Galax, Virginia, which resulted in the death of the two homeowners.[9] The company was charged for violations of FIFRA and fined $500,000, sentenced to serve 24 months of probation and to perform 2,000 hours of community service.[10] Officials indicted Chemical Commodities, located in Olathe, Kansas, after a truck containing hazardous waste caught fire and caused the evacuation of a neighborhood in Kansas City.[11] The company was charged with violations of RCRA and was sentenced to pay a fine of $505,760, a $200 special assessment, and to serve 60 months of probation.[12] Prosecutors charged the Southern Union Company for storing hazardous waste without a permit after three youths broke into the facility and contaminated a nearby apartment complex that had to be evacuated.[13] The company was sentenced to pay a $6 million fine, serve 24 months of probation, and pay $12 million in payments.[14]

Officials charged DLM Farms after employees sprayed a passing school bus and the driver and eight kindergarten age children were exposed to dangerous pesticides.[15] The company was charged with violations of FIFRA and ordered to pay a $2,500 fine. Prosecutors indicted Grenale Leonard Williams and Sunland Pest Control Services for the illegal application of a pesticide that injured a child.[16] Several family members also experienced symptoms of pesticide poisoning.[17] The company was sentenced to serve 5 years of probation and Williams was sentenced to 12 months of incarceration as was co-defendant Canarie Deon Curry.

Government officials charged Terminix, USVI, after the illegal treatment of methyl bromide in a rental unit caused serious injury and hospitalization to the family occupying the unit.[18] For violations of FIFRA, the company was sentenced to pay $9.2 million in criminal fines,

$1,242,250 in restitution, and serve five years of probation. The company will also make a $1 million community service payment.[19] Jose Rivera, the former branch manager for the company, was also indicted in a separate prosecution.[20] He pleaded guilty to four counts of violating FIFRA and was sentenced to serve 12 months in prison.[21] Officials charged Curtis Technology for illegal transport and storage of hazardous materials to three residences in San Diego, California.[22] The area was evacuated and the materials detonated on-site by a bomb squad. The company was charged with violations of RCA and sentenced to pay a $45,000 fine.[23]

Prosecutors indicted the Wade Jones Company for discharging hazardous waste onto the ground that resulted in the evacuation of a local school.[24] The company was charged for violations of RCRA and was sentenced to pay a $50,000 fine and a $200 special assessment. Officials charged Omni Applicators and its president, Mark S. Stewart, for illegally transporting hazardous waste and disposing of pesticides in a manner that injured two county inspectors.[25] The defendants were charged with violations of RCRA and FIFRA and Omni was sentenced to forfeit $60,000 in assets to pay for cleanup and Mark Stewart was sentenced to 12 months and one-day incarceration. Prosecutors indicated VC Tank Lines after a near explosion from stored hazardous waste caused local residents to be evacuated.[26] The company was charged for violations of RCRA and was sentenced to pay a $200 community service fine, pay restitution of $8,838.13 to pay for the evacuation, and pay a $191,162 federal fine. Government officials charged Mount Whitney Pest Control after applying pesticides that was sprayed on 15 high school students, 13 of which required medical attention.[27] The company was charged with violations of FIFRA and was sentenced to pay a $5,000 fine, $125 special assessment, and $10,000 for training programs.

Prosecutors charged Bugman Pest and Lawn for misapplying registered pesticides, which resulted in the death of two children.[28] The company was charged for violations of FIFRA and was sentenced to serve 36 months of probation and pay a $3,000 fine. Officials indicated James Richard Holt, president of the Techsol Chemical Company, after a coal tar spill resulted in 500 people being evacuated from their homes.[29] Holt was charged with violations of RCRA and the CWA for negligently releasing a hazardous substance into a municipal sewer system and was sentenced to serve 60 days of community confinement, six months of home detention, 60 months of probation, and to pay $13,948 in restitution.

Water Pollution Crimes

In eight prosecutions or about 22% of cases under analysis in this chapter, the primary crime at the center of the prosecution focused on water pollution. In all but one prosecution, government officials charged defendants for violations of the CWA and in the other, violations of the Refuse Act. Two of these prosecutions, A-1 Sanitation and CH20, Inc., were discussed previously in the chapter. We provide case examples in this category with the six remaining prosecutions of Equilon Pipeline, Colonial Pipeline Company, Hotels AB, LLC., Prejean-Webb Enterprises, LLC., Seneca Companies, Inc., and Mora Development Company.

Prosecutors indicted Equilon Pipeline and the Olympic Pipeline Company after a rupture in a pipeline caused a release of about 277,000 gallons of gasoline that ignited killing two boys playing nearby.[30] Equilon was sentenced to serve 60 months of probation, pay a $525 special assessment, and pay a $15 million federal fine. Olympic was sentenced to serve 60 months of probation, pay a $650 special assessment, and pay a $6 million fine. Prosecutors indicted the Colonial Pipeline Company after a ruptured pipeline caused 1,257 barrels of diesel fuel to discharge into the Tennessee River in Chattanooga, Tennessee.[31] For violations of the Refuse Act, the company was sentenced to pay a $50,000 federal fine and pay $350,000 in restitution. Prosecutors charged Hotels AB, LLC, after chlorine gas was released into the Los Angeles metro system.[32] Several individuals, including a law enforcement officer, were affected by the fumes. For violations of the CWA, the company was ordered to pay $200,000 in federal fines, $150,000 for community service, and $20,284 to the victims of their crime.

Officials indicted Prejean-Webb Enterprises, LLC., a dry cleaning business, for negligently discharging hazardous waste into the city sewer system in Lafayette, Louisiana.[33] For violations of the CWA, the company was sentenced to pay a $2,500 fine. Prosecutors charged the Seneca Companies, Inc., for negligently releasing hazardous substances into the Duenweg, Missouri, sewer system.[34] The company was sentenced to serve five years of probation and pay a $85,000 fine, as well as $2,707 in restitution. Prosecutors indicted Mora Development Company for discharging sewage from a residential development to the Toa Alta municipal sewer system in Puerto Rico.[35] For violations of the CWA, the company was ordered to pay a $3 million criminal fine, serve five years of probation, and make restitution to a family affected by the criminal conduct.[36]

Air Pollution Crimes

Six prosecutions or roughly 17% of the cases we analyzed in this chapter centered on air pollution crimes. Six of the prosecutions focused on violations of the CAA, where one prosecution involved violations of CERCLA. We previously discussed the prosecutions of Trinity Railcar Repair and Werner Transportation Services. We provide case examples in this category with the remaining five prosecutions of the Surpass Chemical Company, K.S., Inc., and Harcros Chemicals, Inc., and Dyno Nobel, Inc.

Prosecutors charged the Surpass Chemical Company after a tank ruptured at their facility in Albany, New York, that released a cloud of chlorine gas.[37] Numerous individuals were injured and evacuated from the area. The company was charged for violations of the CAA and sentenced to pay $105,000 in federal fines. Prosecutors indicted K.S., Inc., after it released a cloud of anhydrous ammonia into a residential area of Tafuna, American Samoa.[38] A large portion of the village had to be evacuated. For violations of the CAA, the company was sentenced to serve 3 months of probation, pay a $125 special assessment, pay a $40,000 federal fine, make $12,000 in community service payments, and issue a public apology in the local newspaper.

Officials charged Harcros Chemicals, Inc., after the facility released a chlorine gas cloud over Atchison, Kansas, causing an evacuation and injuries to the public, including emergency personnel.[39] The company and co-defendant MGP Ingredients were charged for violations of the CAA and each sentenced to pay a $2 million fine. Prosecutors charged Dyno Nobel, Inc., after the company's plant in St. Helens, Oregon, discharged more than six tons of anhydrous ammonia into the air over a three-day period beginning on July 30, 2015.[40] The discharges caused numerous complaints from victims in the community, including foul odors, eye irritation, and difficulty breathing from residents of nearby Columbia City, Oregon. The company was charged for violating CERCLA for failing to notify officials of the release of a hazardous substance and was sentenced to pay a $250,000 fine and to serve two years of probation.[41]

State-Level Environmental Crimes

In three prosecutions or 8% of cases under analysis in this chapter, defendants were charged for state environmental crimes. Clay-Park Laboratories was prosecuted after the company released fumes that caused the evacuation of a school in the Bronx, New York. Fifty-four of the students and 16 adults sought medical treatment after the exposure.[42] The company was charged with violating state environmental laws and was ordered to pay a $35,250 fine. Michael Joseph Kile, Safety Manager at EXPLO Systems, was charged for his role in the previously mentioned prosecution of the company for causing an evacuation of residents of Doyline, Louisiana near Camp Minden.[43] Kile was sentenced to serve 12 weeks of incarceration, pay a $1,000 fine, and pay $1,000 in restitution to the state police. In connection with the prosecution of an oil company in Athens, Georgia, William Alvis Stafford was prosecuted after 6,000 gallons of waste oil product leaked into the public sewer system and residents in a nearby apartment complex were evacuated.[44] Stafford was sentenced to 6 months of incarceration and 114 months of probation.

Punishment and Significant Prosecutions

In the final section of the analysis in the chapter, we turn to describing punishment patterns at sentencing in prosecutions where companies victimize the public. We begin in Table 4.1 with an overview of total monetary penalties assessed at sentencing to all individual defendants and companies in our analysis, listed in the upper-left quadrant of the table. We find that individuals were collectively sentenced to pay over $35 million in monetary penalties in the prosecutions under consideration. Companies were assessed over $59 million in monetary penalties at sentencing for a grand total of some $95 million in penalties across all prosecutions under analysis in the chapter.

In the upper-right quadrant, we show total probation assessed to all individual defendants and companies in the analysis. Individuals were collectively sentenced to serve 658 months of probation or 55 total years. Companies were sentenced to serve 855 months or 71 years of probation at sentencing. In the lower-left quadrant, we show total years of incarceration assessed to defendants at sentencing in our analysis. We find individuals were assessed 330 months or about 28 total years of incarceration in our analysis. A grand total of 2,450 hours of community service

Table 4.1 Total penalties assessed at sentencing for prosecutions where companies victimized the public

Monetary penalties	**Years of probation**
Individuals: >$35 Million	Individuals: 55
Companies: >$59 Million	Companies: 71
Years of incarceration	**Hours of community service**
28	2,450

Source EPA Summary of Criminal Prosecutions Database

(bottom-right quadrant) was also assessed at sentencing. Two thousand of these hours were assessed in the previously discussed prosecution of Orkin Exterminating Company.[45]

Because these general trends in Table 4.1 can be affected by outliers, we discuss examples of prosecutions resulting in significant penalties below. In Table 4.2, we explore cases of significant probation assessed at sentencing. The most significant probation penalty was assessed in the previously discussed case of Equilon Pipeline, where a pipeline ruptured resulted in a large gasoline spill that ignited resulting in death.[46] All five defendants were collectively sentenced to serve 16 years of probation in the case. The prosecution of five corporate officers for Explo Systems, Inc., that was discussed previously, where the company illegally stored and transported waste from military munitions that exploded at Camp Minden, Louisiana, causing an evacuation and significant property damage, resulted in 14 years of collective probation assessed at sentencing.[47] The previously discussed prosecution of CH20, Inc., for illegally discharging chemical waste to a POTW, injuring city workers, resulted in 10 years of probation assessed to all five defendants in the case.[48] These three prosecutions resulted in 39 years of probation assessed at sentencing. This represents about 30% of total probation assessed in the analysis in this chapter, demonstrating the strong effect of significant prosecutions on overall trends noted in the first section of the analysis.

We turn to Table 4.3, where we explore cases of significant incarceration at sentencing that have the potential to influence overall incarceration patterns in the analysis. The prosecution of corporate officers at Explo Systems was by far the most extensive prosecution that involved incarceration. Defendants were collectively sentenced to serve 19 years in prison.[49] Employees of Sunland Pest Control Services including Grenale Leonard Williams and Canarie Deon Curry were sentenced to serve 24 months or

Table 4.2 Cases of significant probation assessed at sentencing for prosecutions where companies victimized the public

Defendant	*Fiscal year*	*Crime*	*Total probation (years)*
Equilon Pipeline	2003	Water Pollution	16
Explo Systems/Charles Ferris Callihan	2019	Hazardous Substances	13
CH20, Inc	1998	Water Pollution	10

*Numbers are rounded
Source EPA Summary of Criminal Prosecutions Database

two years in prison for the illegal application of a pesticide that injured a child.[50] AAA Pest Protection, Inc., and both Owner William Robles and employee Pierce Long were charged with violations of FIFRA for a number of structural fumigations that put residents at risk.[51] The company was ordered to pay a $35,000 fine, serve five years of probation, and implement an employee training program. The company was also ordered to pay $2,181 in restitution to three victims. Robles was sentenced to serve 12 months of incarceration and pay a $30,000 fine, where Pierce was sentenced to serve 12 months and a day of incarceration and a year of supervised release. These cases resulted in 23 of incarceration assessed at sentencing, making up 82% of incarceration ordered in the analysis in the chapter. The results show an even more significant influence of a few outliers on sentencing trends than with probation totals noted in Table 4.2.

Table 4.3 Cases of significant incarceration assessed at sentencing for prosecutions where companies victimized the public

Defendant	*Fiscal year*	*Crime*	*Total incarceration (years)*
Explo Systems/Charles Ferris Callihan	2019	Hazardous substances	19
Sunland Pest Control Services	2016	Hazardous substances	2
AAA Pest Protection	2018	Hazardous substances	2

*Numbers are rounded
Source EPA Summary of Criminal Prosecutions Database

In Table 4.4, we explore large monetary penalties assessed at sentencing. The Explo Systems prosecution was the most significant, resulting in over $35 million in monetary penalties in our analysis.[52] The prosecution of Equilon Pipeline resulted in over $21 million in monetary penalties assessed at sentencing.[53] The prosecution of the Southern Union Company resulted in over $18 million in monetary penalties assessed at sentencing.[54] These three prosecutions make up over $74 million in monetary penalties. With total penalties at roughly $105 million for the prosecutions under consideration in the chapter, about 70% come from these cases alone, showing, like incarceration, that a few significant prosecutions have a tremendous impact on overall sentencing trends in these cases.

In the final section of our analysis in the chapter, we discuss cases where restitution was awarded at sentencing. As in the previous chapter that explored employee victimization, it is difficult to be exacting with how much restitution was awarded, as the total amount is not always clear in the case summaries, nor whether and how much restitution was actually awarded to victims in these cases. In nine cases, we find evidence in the case summaries of restitution being awarded. We estimate that over $1.6 million in restitution was awarded collectively in these cases. In five of these cases, it is clear the restitution is awarded to environmental crime victims and in one of these prosecutions it is unclear how much restitution was awarded. We list these in Table 4.5.

In two prosecutions, victims are awarded restitution at sentencing when they were victims of hazardous substances crimes. In two prosecutions, victims of water pollution were awarded restitution at sentencing. We find no evidence of restitution awarded to victims in cases of air

Table 4.4 Cases of significant monetary penalties assessed at sentencing for prosecutions where companies victimize the public

Defendant	*Fiscal year*	*Crime*	*Total monetary penalties*
Explo Systems/Charles Ferris Callihan	2019	Hazardous substances	<$35 Million
Equilon Pipeline	2003	Water pollution	<$21 Million
Southern Union Company	2010	Hazardous substances	<$18 Million

*Numbers are rounded
Source EPA Summary of Criminal Prosecutions Database

Table 4.5 Percentage of cases where restitution was awarded to victims for prosecutions with where companies victimized the public

Hazardous substances crimes 2 Prosecution (8% of Cases)	**Water pollution crimes** 2 Prosecutions (5% of Cases)
Air pollution crimes 0 Prosecutions (0% of cases)	**State environmental crimes** 0 Prosecutions (0% of cases)

Source EPA Summary of Criminal Prosecutions Database

pollution crimes or state environmental crimes. For hazardous substances crimes, the prosecution of DLM Farms, Inc., for illegally spraying a school bus full of kindergarteners with pesticide, resulted in the company paying $7,516 in restitution to its victims.[55] AAA Pest Protection paid $2,181 in restitution to victims.[56] For water pollution crimes, the previously mentioned prosecution of Hotels AB, LLC, for releasing toxic gas resulted in $20,284 dollars of restitution awarded to victims.[57] The prosecution of Mora Development Company for illegally discharging into a sewer system resulted in restitution to a family, whose child was injured by the company's criminal conduct, but the case summary does not give an amount.[58] The prosecution of the Seneca Companies resulted in $2,707 in restitution but it is unclear in the summary if this went to any victims specifically in the case.[59]

Conclusion

Our exploration of prosecutions of companies that committed environmental crimes against the public has generated a few important implications. Our first important finding is that prosecuting companies for victimizing the public has been rare over time in the United States. Since 1983, we located 36 prosecutions of companies that committed this category of crime. Prosecutions were rare through the 1980s and then sporadic since, with a slight uptick since 2010. We found prosecutions occurring across 25 U.S. states and territories, with California (4), Louisiana (3), and New York (3) having the most prosecutions, but no particular state or region dominating the analysis. These findings are somewhat similar in terms of the overall distribution of cases over time, as in the previous chapter, but prosecutions are less geographically centered, with 28% occurring within three states.

A second important finding related to criminal behavior and serious consequences stemming from these crimes. We find only 16% of prosecutions contained at least one case of a defendant being charged with Title 18 behaviors, such as false statements or fraud. Yet the public suffered grave consequences when companies committed environmental crimes. In 11% of cases, at least one person died from an environmental crime, in 44% they were injured, in 42% people were injured and/or evacuated, and in three% of prosecutions death and injury resulted from an environmental crime.

A third finding of note is the significance of penalties and other assessments levied at sentencing in our analysis. Some $95 million in monetary penalties were assessed to defendants in our analysis, as well as 126 years of probation and 28 years of incarceration. Yet these trends are heavily affected, particularly monetary penalties and incarceration, by a few significant prosecutions. While it is important prosecutors obtain such noteworthy penalties at sentencing, those cases also tend to drive the larger trends in sentencing, making the median or average penalty significantly reduced.

Notes and References

1. Meg Kinnard, 2013, Georgia Firm Indicted in SC Woman's Ammonia Death, AP, Available from: https://news.yahoo.com/georgia-firm-indicted-sc-womans-205917507.html.
2. Werner Transportation Services, Inc. (D. South Carolina 3:13-CR-97-PJG, 2013).
3. A-1 Sanitation (D. Massachusetts 97 MA 0233, 1998).
4. Harrison Research Laboratory (D. New Jersey 00-CR-58, 2000).

   U.S. EPA, 2000, Woman Admits Testing Pesticides on Unknowing Subjects: Corporation Pleads Guilty to Obstruction of Justice, Available from: https://www.epa.gov/archive/epapages/newsroom_archive/newsreleases/fcd9ec28841394c885257173006d6671.html.
5. Friendly Systems, Inc. (D. South Dakota 99-CR30023-001, 2000).
6. JTA Real Estate Brokerage and Property (D. New Hampshire 01-123-02-M, 2002).

7. U.S. EPA, 2002, New Hampshire Company, Manager Sentenced in Fatal Lead Exposure Case, Available from: https://www.epa.gov/archive/epapages/newsroom_archive/newsreleases/64ad1ea999528c0f85256b920060516f.html.
8. Trinity Railcar Repair, Inc. (N.D. New York 1:02-CR-444-001, 2003).
9. Mary Ellin Arch, 1988, Orkin Held Responsible for 2 Fumigation Deaths, Washington Post, Available from: https://www.washingtonpost.com/archive/local/1988/11/17/orkin-held-responsible-for-2-fumigation-deaths/34fd0939-f0c9-41fc-bb7e-b06dfae1af32/.
10. Orkin Exterminating Company (W.D. Virginia 88-00040, 1989).
11. Chemical Commodities (D. Kansas 90-20002-01, 1990).
12. The company's facilities became part of a Superfund cleanup, see: U.S. EPA, 2024, Superfund Success Stories: EPA Region 7, Available from: https://www.epa.gov/superfund/superfund-success-stories-epa-region-7.
13. Southern Union Company (D. Rhode Island 1:07CR00134-01S, 2010).
14. U.S. Department of Justice, 2009, Southern Union Company is Penalized $18 Million for Illegally Storing Mercury at a Rhode Island Site, Available from: https://www.justice.gov/opa/pr/southern-union-company-penalized-18-million-illegally-storing-mercury-rhode-island-site.
15. DLM Farms, Inc. (E.D. California MJ 00201 SKO, 2012).
16. Sunland Pest Control Services, Inc. (S.D Florida 2:16-CR-14001-JEM, 2016).
17. U.S. Department of Justice, 2016, Fumigation Company and Two Individuals Sentenced in Connection with Illegal Pesticide Application Resulting in Injuries to a Minor, Available from: https://www.justice.gov/opa/pr/fumigation-company-and-two-individuals-sentenced-connection-illegal-pesticide-application.
18. Terminix, USVI (D. Virgin Islands 3:17-CR-00007-CVG-RM, 2018).
19. CBS News, 2016, Terminix Fined $10M in Virgin Islands Toxic Pesticide Case, Available from: https://www.cbsnews.com/news/terminix-fined-10-million-in-virgin-islands-toxic-pesticide-case/.

The company later settled with the victims in the case:

Mat Probasco, 2022, V.I. Terminix Tourist Poisoning Case Settled, Available from: https://stthomassource.com/content/2022/03/03/v-i-terminix-tourist-poisoning-case-settled/.

20. U.S. Department of Justice, 2018, Terminix Virgin Islands Branch Manager Pleads Guilty to Four Counts of Illegally Applying Restricted-Use Pesticide to Multiple Residences in the U.S. Virgin Islands, Available from: https://www.justice.gov/usao-vi/pr/terminix-virgin-islands-branch-manager-pleads-guilty-four-counts-illegally-applying.
21. Jose D. Rivera (D. Virgin Islands 3:18-CR-00011-CVG-RM, 2019).
22. Curtis Technology (S.D. California 20-CR-00715, 2020).
23. U.S. Attorney's Office, 2020, Local Firm Ordered to Pay More Than $150,000 for Hazardous Waste Violations, Available from: https://www.justice.gov/usao-sdca/pr/local-firm-ordered-pay-more-150000-hazardous-waste-violations.
24. Wade Jones Company (W.D. Arkansas CR-94-50023, 1994).
25. Omni Applicators (D. Arizona CR 93 0000, 1994).
26. VC Tank Lines (N.D. Indiana 2:98CR69RL, 1999).
27. Mount Whitney Pest Control (E.D. California 1:06 CR-00172-DLB, 2007).
28. Bugman Pest and Lawn (D. Utah 1:11-CR-00017, 2012).
29. James Richard Holt (S.D. West Virginia 3 09 CR 00231, 2011).
30. Equilon Pipeline (W.D. Washington CR01-338R, 2003).

    Daryl C. McClary, 2003, Olympic Pipeline Accident in Bellingham Kills Three Youths on June 10, 1999, Available from: https://www.historylink.org/File/5468.
31. Colonial Pipeline Company (E.D. Tennessee 3:04-CR-49, 2004).

    EPA settled a number of related civil claims against the company for pipeline spills in several states, including a $34 million civil penalty, the largest a company has paid in EPA history:

    U.S. EPA, 2024, Colonial Pipeline Company Clean Water Act Settlement, Available from: https://www.epa.gov/enforcement/colonial-pipeline-company-clean-water-act-settlement.
32. Hotels AB, LLC. (C.D. California CR 09-01287-GW, 2010).
33. Prejean-Webb Enterprises, LLC (W.D. Louisiana 6:12CR-00162, 2012).
34. Seneca (W.D. Missouri 3:13-CR-05022-MDH, 2015).

35. Mora Development Company (D. Puerto Rico 3:16-CR-00018-01, 2018).
36. U.S. Department of Justice, 2018, Mora Development Sentenced to Pay a Fine of $3 Million For Violation of the Clean Water Act, Available from: https://www.justice.gov/usao-pr/pr/mora-development-sentenced-pay-fine-3-million-violation-clean-water-act.
37. Surpass Chemical Company (N.D. New York 98-CR-352, 1998).
38. K.S., Inc. (D. Hawaii 1:06CR00113, 2006).
39. Harcros Chemicals, Inc. (D. Kansas 19-40021-01/02/-HLT, 2020).
40. Dyno Nobel, Inc. (D. Oregon 3:18-CR-63-SI, 2018).
41. United States' Attorneys Office, 2018, Dyno-Nobel, Inc. Pleads Guilty for Failing to Notify Federal Authorities of Anhydrous Discharges Near St. Helens, Available from: https://www.justice.gov/usao-or/pr/dyno-nobel-inc-pleads-guilty-failing-notify-federal-authorities-anhydrous-discharges-near.
42. Clay-Park Laboratories (Clay-Park Laboratories, Inc., State, 1997).
43. Michael Joseph Kile (Louisiana State Court 87107, 2014).
44. William Alvis Stafford (Georgia State Court SU-05-CR-0917-5, 2014).
45. Orkin Exterminating Company (W.D. Virginia 88-00040, 1989).
46. Equilon Pipeline (W.D. Washington CR01-338R, 2003).
47. Charles Ferris Callihan (W.D. Louisiana 16-CR-00214-06, 2019).
48. CH20, Inc. (W.D. Washington CR-97, 1998).
49. Charles Ferris Callihan (W.D. Louisiana 16-CR-00214-06, 2019).
50. Sunland Pest Control Services, Inc. (S.D. Florida 2:16-CR-14001-JEM, 2016).
51. AAA Pest Protection, Inc. (S.D. Florida 17-60205-CR, 2018).
52. Charles Ferris Callihan (W.D. Louisiana 16-CR-00214-06, 2019).
53. Equilon Pipeline (W.D. Washington CR01-338R, 2003).
54. Southern Union Company (D. Rhode Island 1:07CR00134-01S, 2010).
55. DLM Farms, Inc. (E.D. California MJ 00201 SKO, 2012).
56. AAA Pest Protection, Inc. (S.D. Florida 17-60205-CR, 2018).
57. Hotels AB, LLC (C.D. California CR 09-01287-GW, 2010).
58. Mora Development Company (D. Puerto Rico 3:16-CR-00018-01, 2018).
59. Seneca Companies, Inc. (W.D. Missouri 3:13-CR-05022-MDH, 2015).

CHAPTER 5

When Individuals Victimize the Public

Abstract This chapter describes the ways individuals harm the public and each other in the commission of environmental crimes in the United States. We explore 27 prosecutions where one or more individuals are prosecuted for committing environmental crimes against the public. We find these prosecutions result in injuries, evacuations, and even death. We mirror the flow of the previous two chapters, by exploring patterns in prosecutions over time and place, followed by an exploration of general themes in prosecutions over time, and then a discussion of sentencing and punishment patterns in these prosecutions.

Keywords Public victimization · Individual environmental crimes

Margaret Stewart sold Minnie Lou Rudd's daughter an unmarked milk jug. When Minnie Lou consumed a glass of what she thought was milk, she became ill and died. Investigators found that Stewart illegally sold the registered pesticide, Endosulfan, in unmarked containers to her daughter and other members of the public. She was prosecuted for violations of FIFRA and was sentenced to serve 12 months of incarceration, 12 months of probation, and to pay a $25 fee.[1]

J. Ozymy and M. Jarrell Ozymy, *The U.S. Administrative State and the Protection of Environmental Crime Victims*, Palgrave Studies in Green Criminology, https://doi.org/10.1007/978-3-031-62680-7_5

The public is harmed when individuals engage in serious or chronic violations of environmental law that cause injury, death, and evacuation. Minnie Lou Rudd's death was completely avoidable, had the defendant not engaged in these illegal actions. Our analysis of EPA-CID criminal investigations that lead to criminal prosecution find over two dozen cases where individuals are prosecuted for committing environmental crimes against others and we discuss these below.

Patterns in Prosecutions

In this first section of the analysis, we describe the number of prosecutions of individuals victimizing the public by EPA fiscal year. Figure 5.1 illustrates that two prosecutions are adjudicated in 1998 and three additional prosecutions are adjudicated the following year. One prosecution is adjudicated in 2004 and 2005, two in 2006, and three in 2007. The high point for annual prosecutions occurs in 2011, when five prosecutions are adjudicated with four adjudicated the following year. A grand total of 27 prosecutions were adjudicated, averaging about 0.7 prosecutions annually in our analysis.

In Fig. 5.2, we explore the geography of prosecutions of individuals that victimized the public. Prosecutions occur within 18 U.S. states, with the largest number of prosecutions (4) occurring in Pennsylvania. Three prosecutions were adjudicated within Kansas, Michigan, and New York, respectively. One prosecution was adjudicated within each of the remaining states. Prosecutions averaged 1.5 per state in our analysis.

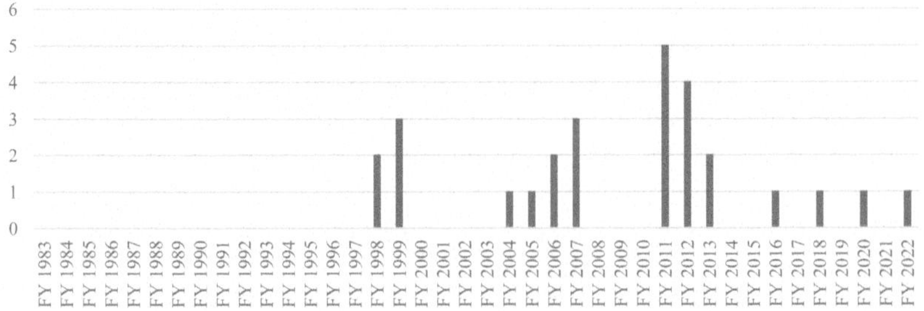

Fig. 5.1 Number of Prosecutions where Individuals Victimized the Public by EPA Fiscal Year, 1983–2022 (*Source* EPA Summary of Criminal Prosecutions Database)

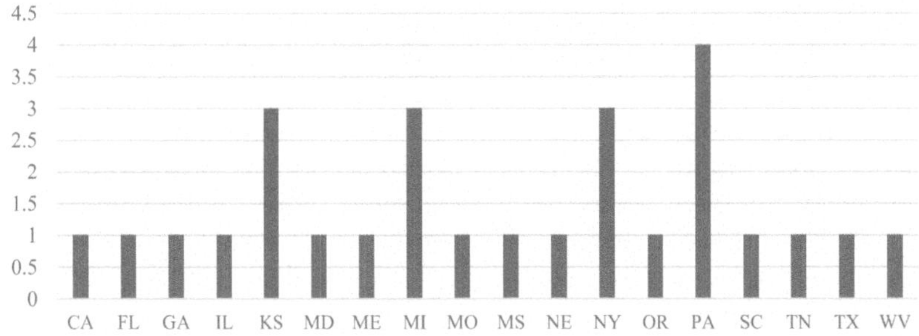

Fig. 5.2 Number of prosecutions where individuals victimized the public by U.S. State and Territory, 1983–2022 (*Source* EPA Summary of Criminal Prosecutions Database)

In Fig. 5.3, we describe charging patterns in prosecutions where individuals victimized the public. TSCA was the most common statute violated in our analysis (8 prosecutions), owning mostly to prosecutions of individuals for lead-based paint crimes. The second most common violation occurred in five prosecutions, where one or more individuals engaged in criminal actions related to pesticides that violated FIFRA. In four prosecutions, at least one individual engaged in a water pollution crime that injured members of the public. In two prosecutions, at least one individual committed an air pollution crime in violation of the CAA. In one prosecution, at least one individual committed a hazardous waste crime in violation of RCRA. In three prosecutions, at least one defendant violated state environmental laws.

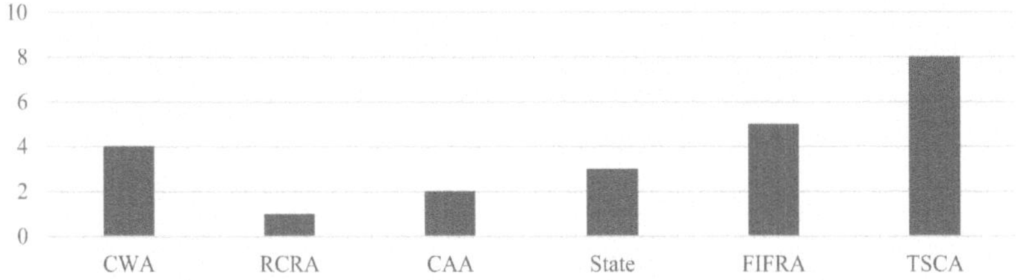

Fig. 5.3 Charging patterns in prosecutions where individuals victimized the public, 1983–2022 (*Source* EPA Summary of Criminal Prosecutions Database)

In Fig. 5.4, we show patterns in criminal behavior in prosecutions where individuals victimized the public. The most common Title 18 violation or other such criminal violation we find in the data involves giving false statements to government officials or on official documents. We find that six prosecutions or about 22% of prosecutions involve giving false statements. In three prosecutions or about 11% of cases in our analysis, defendants were charged with conspiracy. In two prosecutions or roughly 7% of cases in our analysis, at least one defendant was charged with obstruction.

In Fig. 5.5, we describe outcomes in prosecutions where individuals victimized the public. The most common outcome in our analysis was injury. We find that in 21 prosecutions or about 78% of cases in our analysis in the chapter, at least one individual was injured as the result of an environmental crime. In two prosecutions or 7% of cases analyzed in the chapter, at least one individual was killed as the result of an environmental crime. In three cases or about 11% of prosecutions under analysis in the chapter, the public was harmed by an environmental crime that resulted in an evacuation. In one prosecution or about 4% of prosecutions, both injury and evacuation occurred in the case. We provide case examples of the prosecution of Damon Kelly, Shannon Presley, and Edward Louis Wyman.

Prosecutors charged Damon Kelly after he stole pesticide, placed it in soft drink bottles, and sold it to two individuals, one of which unknowingly assumed it was soda, drank it and died. Kelly was charged with state violations including petty larceny and violations of state pesticide control laws. He was sentenced to pay $500 in restitution and a $1,288

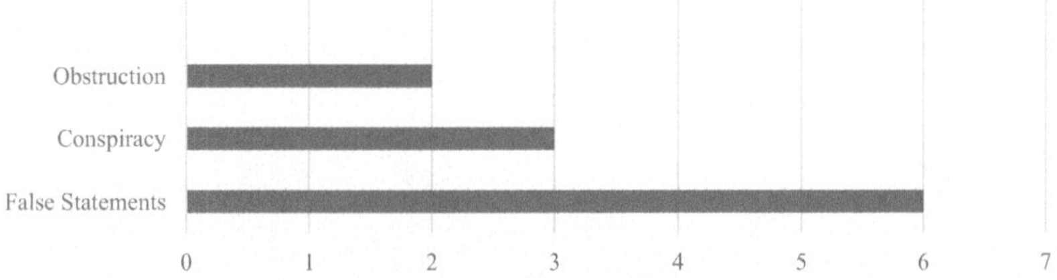

Fig. 5.4 Patterns in criminal behavior in prosecutions where individuals victimized the public, 1983–2022 (*Source* EPA Summary of Criminal Prosecutions Database)

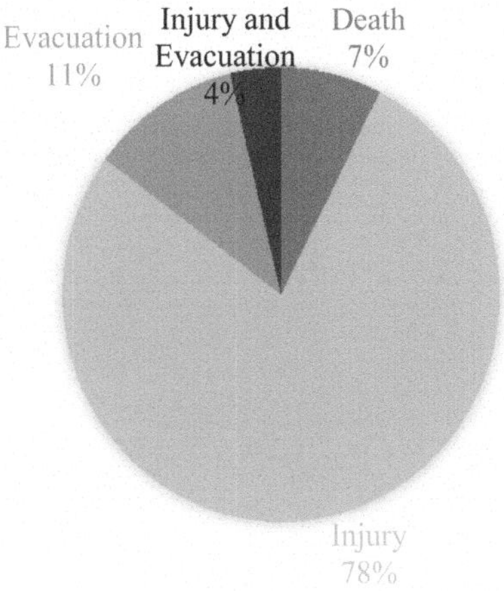

Fig. 5.5 Categories of victimization in prosecutions where individuals victimized the public (*Numbers are rounded. *Source* EPA Summary of Criminal Prosecutions Database)

state fine.[2] Officials indicated Shannon Presley after he applied pesticides to a dwelling and the residents became ill. The defendant was not a certified pesticide applicator and used a pesticide that was prohibited for indoor use. The defendant was charged with one count of aggregated battery and two counts of child endangerment and was sentenced to serve 34 days of incarceration, 18 months of probation, a $5,000 state fine, and $14,043 in restitution to his victims.[3] Government officials charged Edward Louis Wyman after a fire in his residence resulted in an evacuation of his neighborhood.[4] The defendant stored about one million rounds of ammunition, gun powder, and hazardous industrial solvents. He was charged with RCRA violations, including knowing endangerment, and was sentenced to serve 60 months of incarceration, three years of supervised release, and $800,000 to EPA for cleanup costs.

Themes in Prosecutions

We now turn to describing themes in prosecutions where individuals injured the public in the commission of an environmental crime. As in previous chapters, we organized prosecutions around, what we felt in our best judgment was the primary crime at the center of the case. When doing so, we found crimes revolved around similar themes to the previous chapters. We organize these below in order of prevalence in the data,

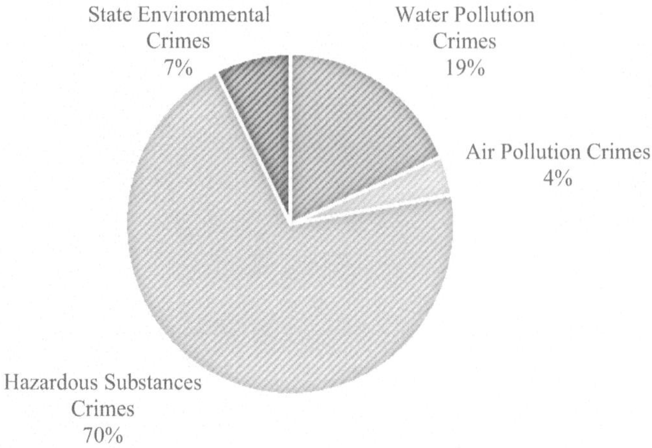

Fig. 5.6 Themes in prosecutions where individuals victimized the public (*Source* EPA Summary of Criminal Prosecutions Database)

beginning with hazardous waste crimes, water pollution crimes, state environmental crimes, and air pollution crimes. The previously mentioned prosecutions of Damon Kelly and Shannon Presley make up all of the state environmental crimes in this category. The only air pollution crime was the prosecution of Eric Kung Shou Ho.[5] The defendant was overseeing renovation work at a hospital and illegally removed asbestos from the facility after a city inspector halted the work. Being done at night, a gas pipeline was opened that caused an explosion that injured three people. Ho was charged for violations of the CAA for the illegal asbestos renovation and was sentenced to serve two months of incarceration, 24 months of probation, to pay a $200 fee, and a $20,000 federal fine (Fig. 5.6).

Hazardous Substances Crimes

In 19 prosecutions or 70% of cases under analysis in the chapter, we find the prosecution centers on crimes involving hazardous substances. These crimes involved pesticides, lead-based paint, and dangerous chemicals. Offenders violated TSCA, FIFRA, CAA, and state environmental laws in this category. We provide case examples with the prosecution of Ruben Brown, Daniel Trehey, Robert Kelly, Jr., Kurvin Grove, Bruce Zauf, Martin Kimber, Maureen Walck, and Dennis Morgan.

Prosecutors charged Ruben Brown for spraying thousands of homes in the Chicago, Illinois area with a restricted use pesticide meant for outdoor use.[6] Some residents had to be relocated and EPA found more than 100 homes had been contaminated and required cleanup. Brown was charged with violations of FIFRA and was sentenced to 24 months of incarceration, and a $50 special assessment fee. In one of the more notable prosecutions in our data, officials indicated Daniel Trehey for misapplying a pesticide in EPA's Region 7 office in Kansas City, Kansas.[7] The third floor of the building was evacuated and several EPA employees became ill and required medical attention. Trehey was charged with violations of FIFRA and was sentenced to serve 12 months of probation and to pay a $2,000 federal fine. Prosecutors indicted Robert Kelly, Jr. for illegally purchasing and applying methyl parathion, a restricted use pesticide that is not labeled for indoor use, in hundreds of homes in the Memphis, Tennessee area.[8] Many were exposed to high levels of the pesticide and became ill after it was applied in their homes. Kelly was charged with 42 counts of violating FIFRA and conspiracy and was sentenced to serve 20 months of incarceration and pay $250,000 in federal fines.

Government officials charged Kurvin Grove after one of the children living in one of his rental properties contracted lead poisoning.[9] Grove was charged with obstruction and was sentenced to serve 12 months of probation, 6 months of home detention, and pay a $100 special assessment fee and a $7,500 fine. Officials charged Bruce Zauf after he failed to disclose the presence of lead-based paint in 11 of his rental units and four children experienced lead poisoning that lived in two of his properties.[10] Zauf was charged with violating TSCA and sentenced to serve 36 months of probation, perform 100 hours of community service, pay a $25 special assessment fee and a $500 federal fine.

In another notable case, government officials indicated Martin Kimber after he engaged in a bizarre revenge plot against the Albany Medical Center in Albany, New York.[11] In retaliation for a billing dispute, Kimber spread mercury all around the facility, including bathrooms, hallways, and the cafeteria, where he placed the hazardous substance in the coffee station, salad bar, and other food stuffs. Kimber plead guilty to using a chemical as a weapon and tampering with consumer products. He was sentenced to serve 168 months of incarceration, 60 months of supervised release, forfeit his home and car, and pay $200,451 in restitution to the Albany Medical Center and pay restitution to his victims.

Prosecutors charged Maureen Walck, a real estate broker, for failing to disclose known lead-based paint hazards in a home she placed under contract.[12] Walck disclosed the lead-based paint hazards to a previously interested buyer, but did not make the known disclosures the final buyer of the home, whose child was sick with lead poisoning. Walck was sentenced to time served, fined $1,000, and to pay $53,326.07 in restitution to the victim. Officials charged Dennis Morgan for failing to provide lead-based paint hazards in one of his rental properties in Sunbury, Pennsylvania.[13] A child occupying the unit tested positive for lead poisoning. Morgan was charged with violations of the TSCA and was sentenced to pay a $2,000 fine and serve one year of probation.

Nine additional prosecutions occurred in this category, include the prosecution of Paul H. Bowman, Thomas Walz, Arnoldo Carrillo Bazan, MoReno Rondell Taylor, Donald Mickey Patterson, Anthony Sharpe, Ji Hui Yang, Cephus Moses Murrell, and Martin Glaves Kuna. Officials charged Paul H. Bowman for failing to disclose lead in a rental property and a 14-month-old tenant was diagnosed with high levels of lead in their blood.[14] Bowman was sentenced to serve 12 months of probation, 3 months of home detention, pay a $100 special assessment and a $200 federal fine. Prosecutors indicated Thomas Walz after he sold a home and indicated no presence of lead-based paint, when he knew otherwise, and a young child was diagnosed with elevated levels of lead in their blood.[15] Walz was charged with fraud and making false statements and was sentenced to serve 36 months of probation, to pay a $25 special assessment fee, and a $5,000 federal fine. Government officials indicated Arnoldo Carrillo Bazan for poisoning patrons at Mi Ranchito restaurant in Olathe, Kansas.[16] Bazan poisoned the salsa bar with insecticide and patrons became ill. Bazan was sentenced to serve 120 months of incarceration for tampering with a consumer product, as well as 24 months of supervised release, to pay $248,673 in restitution to the restaurant, and $487,763.08 in restitution to his victims. Prosecutors indicated MoReno Rondell Taylor for falsifying lead-based paint testing.[17] He was sentenced to serve 120 days of home confinement and 24 months of probation for making false statements and TSCA violations.

Officials charged Donald Mickey Patterson for failing to properly test for lead-based paint, when a child was admitted to the hospital with high levels of lead in their blood in the residence.[18] For fraud, false statements, and violations of TSCA, Patterson was sentenced to serve 46 months of incarceration, 24 months of probation, and $662 in restitution to his victims. Prosecutors indicated Anthony Sharpe for falsifying lead-based paint sampling in conjunction with the prosecution of MoReno Taylor.[19] For making false statements, mail fraud, and transportation of stolen goods, Sharpe was sentenced to 18 months of incarceration, 24 months of probation, ordered to pay $1,600 in restitution.

Government prosecutors indicated Ji Hui Yang and 11 co-defendants for illegally selling registered pesticides in Washington Heights, Corona, and Flushing Queens, New York.[20] An individual became ill after ingesting the mislabeled pesticide. The defendants sold some 4,500 packages of the unregistered pesticide to undercover agents. The defendants were collectively sentenced to serve 24 months of probation and pay over $29,000 in fines and restitution. Prosecutors charged Cephus Moses Murrell owned a number of rental properties in Baltimore, Maryland.[21] After children in the units reported high blood levels, inspectors conducted investigations and confirmed the lead poisoning despite having warned the defendant in the past. The defendant was sentenced to serve 12 months and a day of incarceration, 12 months of supervised release, and six months of home detention. Officials indicated Martin Glaves Kuna for fraudulently conducting lead-based paint inspections.[22] Kuna failed to perform the tests in a home where children experienced elevated levels of lead in their blood. Kuna plead guilty to wire fraud and was sentenced to serve 14 months of incarceration, and to pay $2,372 in restitution to his victims.

Water Pollution Crimes

In five prosecutions or 19% of cases under consideration in our analysis, defendants were charged with water pollution crimes. All but one of the prosecutions involved charging defendants with violations of the CWA. We give case examples with the prosecution of Allen Frey, David Eugene Turner, Robert Gunter, Daryl D. Quinn, and Jaron Coleman.

Prosecutors charged Allen Frey for discharging pollutants into a POTW. Fumes injured workers in the plant causing them to become ill with headaches and nausea.[23] Frey was charged with violations of the CWA and was sentenced to pay a $5,000 fine. Turner and seven co-defendants were charged with a number of crimes, including violations of the CWA for illegally discharging chemicals into a creek.[24] Prosecutors indicate Robert Gunter directed high school students to illegally dispose of lab supplies at Big Creek High School in War, West Virginia.[25] The students developed nose bleeds and other symptoms, one went to the hospital, and the school was closed the following day. Gunter was prosecuted for violations of the CWA for the illegal discharge of chemicals and was sentenced to pay a $2,500 federal fine and a $25 special assessment fee. Officials charged Daryl D. Quinn, a salesman for Water Treatment Equipment, Incorporated, after they installed a water purification system meant to remove uranium from a customer's water well, but Quinn lied to the customer, as they had been exposed to unsafe levels of uranium.[26] Quinn was charged with mail fraud, sentenced to serve 36 months of probation, ordered to pay a $3,500 fine, and $950 in restitution to the victims. Government officials charged Jaron Coleman for dumping 3,000 gallons of gasoline on the ground at a gas station in Thomasville, Georgia.[27] The fuel discharged into a creek and caused the evacuation of a nearby school as it migrated to ditches around the facility. Coleman was charged with violations of the CWA and was sentenced to serve 18 months in prison.

Punishment and Significant Prosecutions

In the final section of the analysis in the chapter, we turn to discussing punishment and sentencing patterns in cases where individuals victimized the public. We begin with a discussion of overall trends in punishment, followed by a description of significant prosecutions that influence these figures, and then end with an overview of restitution ordered at sentencing in these prosecutions.

In Table 5.1 we explore total penalties assessed at sentencing for prosecutions where individuals victimized the public. In the upper-left quadrant, we show that individual defendants were collectively sentenced to pay over $4.8 million in monetary penalties in our analysis. Companies were sentenced to pay $5,000 in monetary penalties.[28] In the upper-right quadrant, individuals were collectively sentenced to serve 68 years of probation in our analysis. In the lower-left quadrant, we show that defendants were collectively sentenced to serve 103 years of incarceration and in the lower-right quadrant, 100 hours of community service was assessed to defendants at sentencing in our analysis.

Table 5.1 Total penalties assessed at sentencing for prosecutions where individuals victimized the public

Monetary penalties Individuals: >4 Million	**Years of probation** Individuals: 68
Years of incarceration 103	**Hours of community service** 100

Source EPA Summary of Criminal Prosecutions Database

In Table 5.2, we begin to describe significant prosecutions that influence probation outcomes in the analysis. The prosecution of David Eugene Turner was by far the most significant prosecution in terms of probation outcomes. Turner and his co-defendants, including Robert Evans the boss of the labor camp, were collectively sentenced to 324 months of 27 years of probation for running an illegal, forced labor camp, where homeless individuals were paid in drugs and untaxed cigarettes and alcohol.[29] Arnoldo Carrillo Bazan and his co-defendant Yini De La Torre, were collectively sentenced to five years of probation for a revenge poisoning of patrons at their former place of employment that poisoned 36 patrons of the Mi Ranchito restaurant.[30] For his role in carrying out his own revenge plot against a hospital to which he had a billing dispute that involved contaminating the facility with mercury, Martin Kimber was sentenced to serve five years of probation.[31] These three cases cumulatively resulted in 37 years of probation assessed at sentencing. Total probation assessed to individual defendants was 68 years, making these cases responsible for 54% of total probation assessed at sentencing in our analysis, showing the strong effect of a few significant prosecutions on overall sentencing patterns in the data.

Table 5.2 Cases of significant probation assessed at sentencing for prosecutions where individuals victimized the public

Defendant	*Fiscal year*	*Crime*	*Total probation (years)*
David Eugene Turner	2007	Water pollution	27
Arnoldo Carrillo Bazan	2011	Hazardous substances	5
Martin Kimber	2013	Hazardous substances	5

*Numbers are rounded
Source EPA Summary of Criminal Prosecutions Database

In Table 5.3, we explore cases of significant incarceration affecting broader patterns in sentencing in our analysis in this chapter. The prosecution of David Eugene Turner resulted in over 52 years of incarceration cumulatively assessed to defendants at sentencing. The prosecution of Arnoldo Carrillo Bazan resulted in 17 years of incarceration at sentencing. Martin Kimber was sentenced to serve 14 years of incarceration. These three cases resulted in 83 years of incarceration at sentencing. With 103 years of cumulative incarceration assessed to defendants at sentencing in our analysis in the chapter, these cases make up about 81% of total incarceration in the analysis. Like probation totals, a few significant prosecutions have a profoundly substantive impact on overall trends in the analysis.

In Table 5.4, we explore cases of significant monetary penalties assessed to defendants at sentencing in our analysis. The most important prosecution here as again against Robert Evans, Sr., David Eugene Turner, and their co-defendants. Fines, fees, and judgments against the defendants resulted in over $2.2 million in monetary penalties. Restitution and fees levied at sentencing against Arnoldo Carrillo Bazan resulted in over $1.2 million in monetary penalties ordered at sentencing. The prosecution over Edward Louis Wyman resulted in the court ordering the defendant to pay $800,000 to EPA to reimburse the agency for the cost of the 47-day cleanup associated with his illegal ammunition, gun powder, and chemical stash.[32] These cases resulted in over $4.2 million in monetary penalties levied at sentencing. With over $4.8 million in monetary penalties assessed to defendants in our analysis in the chapter, about 87% can be

Table 5.3 Cases of significant incarceration assessed at sentencing for prosecutions where individuals victimized the public

Defendant	*Fiscal year*	*Crime*	*Total incarceration (years)*
David Eugene Turner	2007	Water pollution	52
Arnoldo Carrillo Bazan	2011	Hazardous substances	17
Martin Kimber	2013	Hazardous substances	14

*Numbers are rounded

Source EPA Summary of Criminal Prosecutions Database

Table 5.4 Cases of significant monetary penalties assessed at sentencing for prosecutions where individuals victimized the public

Defendant	*Fiscal year*	*Crime*	*Total monetary penalties*
David Eugene Turner	2007	Water pollution	<$2.25 Million
Arnoldo Carrillo Bazan	2011	Hazardous substances	<$1.2 Million
Edward Louis Wyman	2012	Hazardous substances	$800,000

*Numbers are rounded
Source EPA Summary of Criminal Prosecutions Database

attributed to these three cases, continuing the previous trends found with significant prosecutions impacting probation and incarceration totals.

We now turn to the issue of restitution ordered to victims in the analysis in Table 5.5. As in previous chapters, it is not always clear from the case studies how much restitution is paid and if it is paid to victims or other government or private entities for the cost of cleanup or damage. We find 10 prosecutions where restitution was ordered, meaning in about 37% of prosecutions in the chapter defendants were ordered to pay restitution in the case. Seven of these prosecutions occurred in hazardous substances prosecutions, one case of restitution in a water pollution prosecution, two cases in prosecutions of state environmental crimes, and no cases in air pollution prosecutions in the chapter. We estimate that about $1,311,569 in total restitution was paid in these cases, with the amount unclear in one prosecution. Our estimate of restitution paid directly to victims in the prosecutions analyzed in this chapter is $1,047,796.03, with 93% of that total attributed to the Arnoldo Carrillo Bazan prosecution, spread across eight prosecutions where it was clear restitution had been paid to victims.

Table 5.5 Percentage of cases where restitution was awarded to victims for prosecutions with where individuals victimized the public

Hazardous substances crimes 7 Prosecutions (26% of cases)	**Water pollution crimes** 1 Prosecutions (4% of cases)
Air pollution crimes 0 Prosecutions (0% of cases)	**State environmental crimes** 2 Prosecutions (7% of cases)

Source EPA Summary of Criminal Prosecutions Database

Conclusion

A few noteworthy implications flow from the results of the analysis in this chapter. First, we find that prosecuting individuals for victimizing the public is extremely rare in the United States. Beginning in 1983, we describe 27 prosecutions that fall properly within this category of victimization. We find few in the 1980s, with 70% occurring during and after 2004 in our analysis. Prosecutions are concentrated in a few states. Cases are adjudicated within 18 U.S. states, with Pennsylvania, New York, Michigan, and Kansas making up 13 prosecutions or 48% of the total cases under analysis in this chapter. Geographic concentration is greater in this chapter than the previous chapters focusing on victimization in the workplace and from companies victimizing the public.

We also find a significant number of cases involve serious criminal behaviors, including Title 18 violations, such as obstruction, conspiracy, and false statements. About 56% of prosecutions analyzed in the chapter contain one or more of these charges. Victims also suffered serious repercussions from the actions of these offenders. In 78% of prosecutions, at least one individual was injured during the commission of an environmental crime, in 11% victims were evacuation, in 4% of cases at least one individual was injured or evacuation, and in 7% of prosecutions, individuals died as the result of an environmental crime.

A final finding of significance is that prosecutors were able to obtain robust penalties at sentencing in these prosecutions, including restitution ordered at sentencing. Over 68 years of probation, 103 years of incarceration, and $4.8 million in monetary penalties were assessed at sentencing in our analysis within this chapter. While these represent significant outcomes to punish and deter environmental offenders, broader patterns are heavily influenced by a few cases. Fifty-four percent of probation, 81% of incarceration, and 87% of monetary penalties can be ascribed to three cases in each category. Restitution occurred in 37% of cases, with about 30% of cases showing direct restitution paid to victims at sentencing. It is also worth mentioning that two prosecutions in particular, that of Ronald Evans, Sr. and Arnoldo Carrillo Bazan, have a particular and significant influence on sentencing outcomes. In both of these prosecutions, the environmental crimes were also secondary to the larger criminal offenses. This was also true in the prosecution of Martin Kimber.

Notes and References

1. Margaret Stewart (N.D. Mississippi 2:98CR012, 1998).
2. Damon Kelly (South Carolina 15:GS-08-0585, 2016).
3. Shannon Presley (Kansas state court-11CR119 PA, 2012).
4. Edward Louis Wyman (C.D. California 2009-CR-577(A)-GHK, 2012).
5. Eric Kung Shou Ho (S.D. Texas H-00-183, 2006).
6. Ruben Brown (N.D. Illinois 97-CR-277-1, 1998).
7. Daniel Trehey (D. Kansas 98-20074-01-GLR, 1999).
8. Robert Kelly, Jr. (W.D. Tennessee 2:98CR20018-001, 1999).
9. Kurvin Grove (M.D. Pennsylvania 1:CR-03-00060-001, 2004).
10. Bruce Zauf (E.D. Missouri 4:05CR530 RWS, 2006).
11. Martin Kimber (N.D. New York 12-CR-506(LEK), 2013).
12. Maureen Walck (W.D. New York 17-MJ-1103, 2018).
13. Dennis Morgan (M.D. Pennsylvania 4:20-CR-00057, 2022).
14. Paul H. Bowman (M.D. Pennsylvania 1:04-CR-395m 2005).
15. Thomas Walz (D. Nebraska 8:06CR200-001, 2007).
16. Arnoldo Carrillo Bazan (D. Kansas 2:09-CR-20135-KHV, 2011).
17. MoReno Rondell Taylor (E.D. Michigan 2:10-CR-20584-NGE-RSW, 2011).
18. Donald Mickey Patterson (E.D. Michigan 10CR20247, 2011).
19. Anthony Sharpe (E.D. Michigan2:10-CR-20584-NGE-RSW, 2011).
20. Ji Hui Yang (New York CR-00-000, 2012).
21. Cephus Moses Murrell (D. Maryland L-11-03330, 2012).
22. Martin Glaves Kuna (D. Oregon 313-CR-0050 SI, 2013).
23. Allen Frey (E.D. Pennsylvania CR:2:99-0497M-001, 1999).

24. David Eugene Turner (M.D. Florida 3:05-CR-00159-TJC-MMH, 2007).
25. Robert Gunter (S.D. West Virginia 1:06-00240, 2007).
26. Daryl D. Quinn (D. Maine 10-CR-106-PH, 2011).
27. Jaron Coleman (M.D. Georgia 7:19-CR-00043, 2020).
28. In the previously discussed prosecution of Anthony Sharpe (E.D. Michigan 2:10-CR-20584-NGE-RSW, 2011), and MoReno Taylor MoReno Rondell Taylor (E.D. Michigan 2:10-CR-20584-NGE-RSW, 2011), Sharpe's company, Sharpe Environmental was named as a co-defendant in the case and sentenced to serve 60 months of probation and pay a $5,000 fine. Because Sharpe was the principal defendant and received the bulk of the punishment, we decided to place this prosecution in this chapter, rather than the chapter focusing on companies injuring the public.
29. David Eugene Turner (M.D. Florida 3:05-CR-00159-TJC-MMH, 2007).

    CBS News, 2005, Feds Raid Florida Labor Camp, Available from: https://www.cbsnews.com/news/feds-raid-florida-labor-camp/.
30. Arnoldo Carrillo Bazan (D. Kansas 2:09-CR-20135-KHV, 2011).

    U.S. Immigration and Customs Enforcement, 2011, Disgruntled Former Restaurant Worker Pleads Guilty to Poisoning Salsa, Available from: https://www.ice.gov/news/releases/disgruntled-former-restaurant-worker-pleads-guilty-poisoning-salsa.
31. Martin Kimber (N.D. New York 12-CR-506(LEK), 2013).

    United States Attorney's Office, 2013, Kimber Sentenced to 14 Years in Prison for Chemical Weapon Attack at Albany Medical Center, Available from: https://www.justice.gov/usao-ndny/pr/kimber-sentenced-14-years-prison-chemical-weapon-attack-albany-medical-center.
32. Edward Louis Wyman (C.D. California 2009-CR-577(A)-GHK, 2012).

    CBS Los Angeles, 2011, Man Convicted of Storing Explosives in His Reseda Home, Available from: https://www.cbsnews.com/losangeles/news/man-convicted-of-storing-explosives-in-his-reseda-home/.

CHAPTER 6

Toward a Victimology of Environmental Crime Victims

Abstract In this chapter, we build on the results of the three previous analysis chapters by exploring the broader universe of victimization within federal environmental crime prosecutions and develop a victimology of environmental crime victims. We begin by creating a typology of environmental crime victims that builds from and dovetails with the previous chapters. We then describe the larger themes that emerge from identifying the primary categories of crimes in these prosecutions.

Keywords Environmental victimology · Typology of environmental crimes · Punishing offenders

We have found in the previous three chapters that the most common forms of environment offending tend to involve employees as well as companies and individuals committing environmental crimes against the public. Punishments for offending can be significant, but serious penalties are rare. While restitution to victims may have been more common than previously considered, it is not routinely very significant.

Building on the analysis of the previous chapters, we typologize environmental crime victims in this chapter. We follow this analysis with a

J. Ozymy and M. Jarrell Ozymy, *The U.S. Administrative State and the Protection of Environmental Crime Victims*, Palgrave Studies in Green Criminology, https://doi.org/10.1007/978-3-031-62680-7_6

discussion of charging themes in prosecutions, common offender behaviors, and categorize the consequences of victimization. We conclude with a description of overall penalties levied at sentencing, including a discussion of restitution awarded to victims at sentencing.

Typologizing Victims of Environmental Crime

The totality of the analysis from the previous three chapters now allows us to build a typology of environmental crime victims. In Fig. 6.1, we show that environmental crime victims fall within four categories. The most common victim type is an employee that is victimized on the job. In 42 prosecutions or 36% of prosecutions analyzed, employees are crime victims. The second most common victim type is when one or more individuals are victimized by a company. In 36 prosecutions or 31% of cases in the analysis, one or more companies victimize at least one individual or the general public. The third most common victim type is one or more individuals that are victimized by another individual. In 27 prosecutions or 23% of cases in the analysis, one or more members of the public are victimized by at least one other offender.

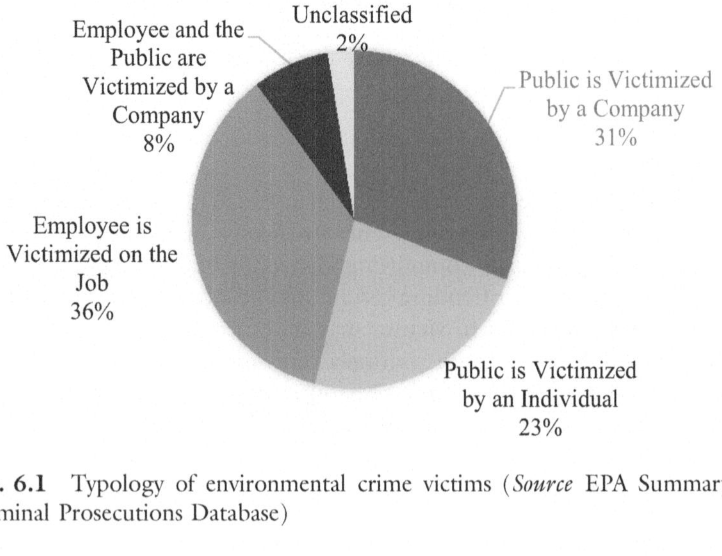

Fig. 6.1 Typology of environmental crime victims (*Source* EPA Summary of Criminal Prosecutions Database)

The most common environmental crime victim is thus, in the following order: an employee, victim of a company, or victim of another individual. In nine prosecutions, an employee and at least one member of the public are victimized by a company, constituting 8% of prosecutions in the analysis. Substantively, it is important to note an important caveat here, as eight of these prosecutions stem from the previously discussed prosecution of the SCWWC.[1] The prosecution of the Airosol Company resulted in injuries to workers and an evacuation of the nearby community as is included in this category as well.[2] These facts do not suggest shifting these prosecutions from a unique category, just the need to clarify that most stem from one case.

Three prosecutions defy categorization in Fig. 6.1. Cecil Morgan and William Gordon Hines, deputies in the Calcasieu Parish Sheriff's Department in Louisiana, were prosecuted for moving chemicals and paraphernalia from a drug lab to a storage shed and then illegally dumped on the Parish's property.[3] A chemical cloud appeared when the materials were dumped and the inmates used to move and dump the material were injured. Each defendant was charged for violations of the CAA and was sentenced to serve six months of probation and each pay a $6,000 fine. In this case, the offenders victimized the inmates in their official capacity, but did not employee them so they were not employees, although they were injured while working on their behalf. We could have included this in the category of the public victimized by individuals, but chose to leave it unclassified.

The second unclassified case was the prosecution of David Lynn Jensen.[4] Jensen injured himself and previously his daughter, by manufactuing explosives in his home that ignited, causing injuries to his hands—he also dumped chemicals down the sink. Charged with reckless endangerment, Jensen was sentenced to serve 60 months of incarceration, 36 months of probation, and to perform 100 hours of community service. The defendant injured himself and his daughter, making the case unique in our data and unable to be categorized in Fig. 6.1.

The final unclassified case involves the prosecution of Steve Eiffler.[5] The defendant engaged in an ill-advised plan to heat mercury in an attempt to process what he thought was gold, from black sand he dumped into the bathtub in his motel room, the Overnight Motor Inn located in Salt Lake City, Utah. Eiffler passed out from the fumes and required hospitalization. Charged with unlawful discharge of pollutants under state law, Eiffler was convicted and sentenced to serve 60 months of incarceration, to pay a $5,000 state fine, and $25,825 in restitution to EPA and the Salt Lake County Health Department. For injuring himself in the commission of an environmental crime, Eiffler deserves his own category, but will be placed as unclassified in the analysis.

How They Were Victimized

Figure 6.2 parcels out each of these prosecutions across the central theme in the prosecution. Half of prosecutions center on hazardous substances crimes involving chemicals, pesticides, hazardous waste, and lead-based paint. The 59 prosecutions in this category focus on violations of RCRA, TSCA, FIFRA, and HMTA. Crimes, such as illegal storage, transport, or disposal of hazardous waste, exposure to dangerous pesticides, mislabeled drums that explode, lead poisoning from consuming lead-based paint, fraudulent testing, and mining injuries were all ways in which people were victimized in this category.

Nineteen percent of prosecutions focus on water pollution crimes. In these 22 prosecutions, common violations primarily occurred under the CWA, but a few cases were prosecuted under the Refuse Act.

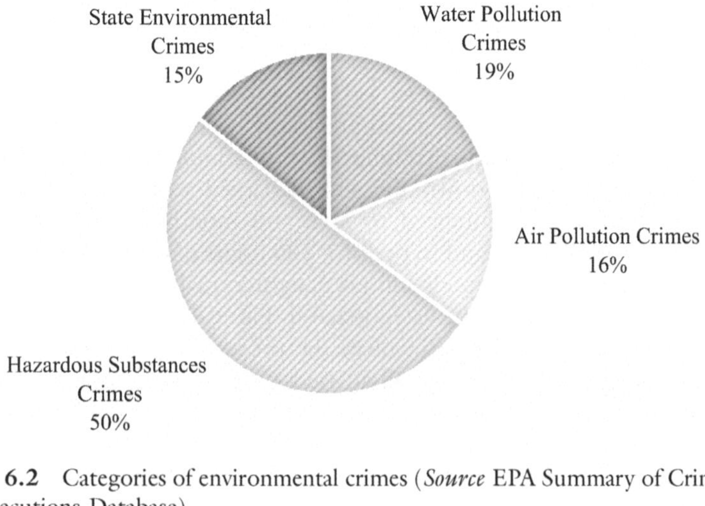

Fig. 6.2 Categories of environmental crimes (*Source* EPA Summary of Criminal Prosecutions Database)

Washing toxic solvents and chemicals down a drain or dumping them in a sewer, ruptured pipelines, operating a forced labor camp that discharged chemicals from drug manufacturing, and causing explosions on offshore oil platforms or discharging oil, were all paths to victimize individuals through water pollution crimes.

Sixteen percent of prosecutions center on air pollution crimes. These 19 prosecutions typically involve violations of the CAA and CERCLA, mostly involving the illegal discharge of air emissions into the ambient air or illegal abatement and disposal of asbestos. Ten of these prosecutions involve employees injured or killed at work. Explosions at a refinery, leaking tanks that caught fire, and mislabeling cylinders full of toxic gases that explode or injure workers, were common causes of victimization in this category.

Fifteen percent of prosecutions are categorized as state-level environmental offenses. In 17 prosecutions, generally involving harmful chemicals or pesticides, defendants were charged with an array of state environmental violations. Releasing toxic fumes, applying pesticides indoors improperly, injuries sustained at a recycling plant, and fires at a hazardous waste processing facility are all paths to victimization in this category. Forty-seven percent of these prosecutions result from the prosecution of officials of the SCWWC.

Given the nature of these crimes described in the previous figures, the results shown in Fig. 6.3 are not surprising, with 21 CWA prosecutions in our data. At 21 prosecutions, air pollution prosecutions under the CAA made up the second-largest grouping of cases in our analysis. Prosecutions for RCRA violations occurred in 18 cases in our data and both pesticide crimes that violated FIFRA and state environmental crimes, make up 17 cases respectively in the analysis. Nine prosecutions under TSCA and two prosecutions under CERCLA round out charging patterns in our analysis.

In Fig. 6.4, we show the most common offender behaviors that lead to an environmental crime in our analysis. We reanalyzed each case, looking for the most common types of crimes we could pinpoint in the 117 prosecutions under analysis. Our work revealed three of the most prevalent types of crimes, including explosions, pesticide crimes, and lead-based paint crimes.

The most common offender behavior in our data involved an explosion. We found an explosion resulted in victimization in 36 prosecutions or 31% of the prosecutions in our analysis. This offense category was very broad, but most frequently includes exploding containers, explosions

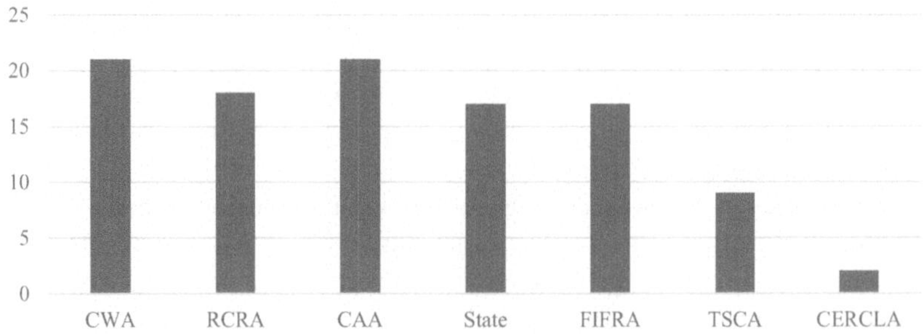

Fig. 6.3 Charging patterns in prosecutions with environmental crime victims (*Source* EPA Summary of Criminal Prosecutions Database)

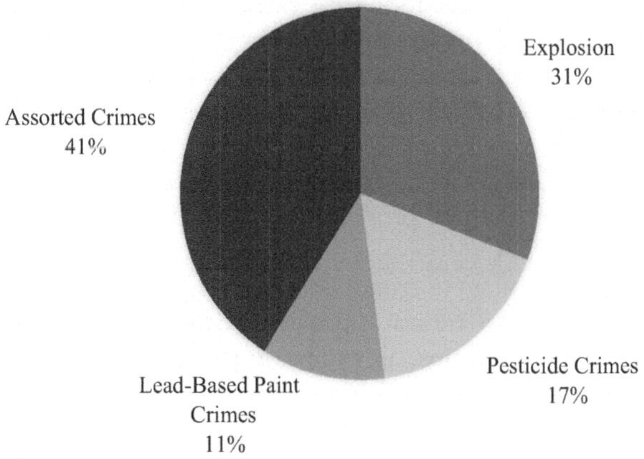

Fig. 6.4 Most common offender behaviors (*Source* EPA Summary of Criminal Prosecutions Database)

at refineries, on offshore oil platforms, and at other industrial facilities. The bulk of the prosecutions (20) involve employees, who were injured at work and 29 involve employees and/or the public, showing workers injured during explosions are the most common single source of victims in our analysis. Nine cases resulted in injury, five in death, 12 in death and injury, six in injury and evacuation, and four in evacuation.

In 20 prosecutions or 17% of the prosecutions in our analysis, the offenders committed crimes involving pesticides. These include misapplication of registered pesticides in residences, illegal commerce in registered pesticides, and applying pesticides in agricultural operations that injure the public. Four of these prosecutions resulted in death, 14 resulted in injury, one prosecution resulted in evacuation, and one lead to evacuation and injury.

Thirteen prosecutions or 11% of cases in our analysis, focused on crimes dealing with lead-based paint. Eleven of these prosecutions involve lead-based paint in the home, one involved falsified lead testing, and one cutting steal beams with lead-based paint, all of which resulted in injury, and one, the prosecution of JTA Real Estate, for failing to provide a family a lead-based paint disclosure, resulted in the death of a child.

The 48 remaining prosecutions, or 41% of prosecutions in our analysis, involve a number of behaviors that are too numerous to parcel into distinct and frequent forms of offending. Most either involve illegal discharges under the CWA, release of hazardous air emissions in violation of the CAA, or hazardous waste violations under RCRA, and two involve mining accidents. Employees were victims in 20 cases. There were 20 cases where companies injured the public, six cases where individuals injured others, and the two unclassified prosecutions of Cecil Morgan for making inmates dispose of hazardous materials, and Steve Eiffler for injuring himself, by melting mercury in his motel room.

In Fig. 6.5, we show the common categories laying out the type of victimization we see in our full data. At 62 prosecutions or 53% of cases in our analysis, one or more people are injured as the result of an environmental crime. Employees were the most common category of victim that was injured during the commission of an environmental crime. We find that 19 prosecutions or about 31% of cases in the Injury category in Fig. 6.5 were employees. Twenty-six percent of prosecutions or 16 cases in this category involve companies injuring the public. In 21 prosecutions or about 34% of prosecutions, individuals were injured by others outside the workplace.

Death and injury occur in 13% of prosecutions, or 15 cases in our data. All but one of these prosecutions involve employees injured or killed on the job. Evacuation also occurs in 15 prosecutions or 13% of our data. All but three of these prosecutions involve companies injuring the public. Evacuation and/or injury occurs in 11 cases or 9% of prosecutions in our analysis. Employees were the most common victims. Finally, in 12% of

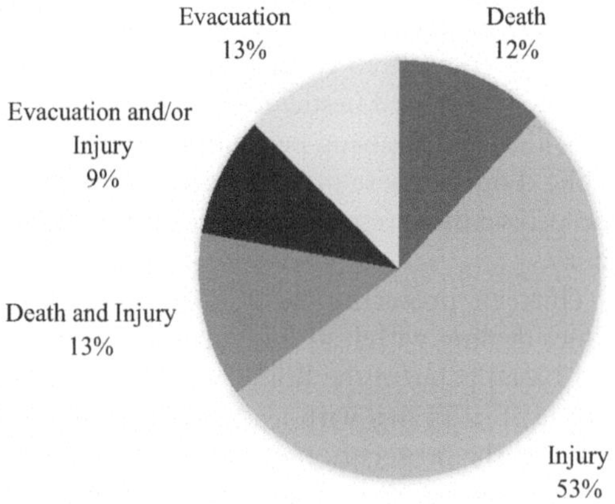

Fig. 6.5 The consequences of victimization (*Source* EPA Summary of Criminal Prosecutions Database)

cases or 14 prosecutions, one or more individuals died from the commission of an environmental crime. Eight of these 14 prosecutions involve employees victimized on the job.

Punishing Environmental Offenders

For the range of behaviors we have listed that injured, killed, or caused people to be evacuated, punishments, we list the punishments in aggregate in Table 6.1. In the upper-left quadrant, we show that monetary penalties exceeded $640 million in our analysis of prosecutions involving identifiable victims. In the upper-right quadrant, we describe 371 years of probation assessed to defendants at sentencing. A total of 198 years of incarceration were assessed to defendants at sentencing and shown in the lower-right quadrant of the table. Finally, in the lower-left quadrant, we show 2,860 hours of community service assessed at sentencing.

Table 6.1 Total penalties assessed at sentencing for prosecutions with environmental crime victims

Monetary penalties	**Years of probation**
>$640 Million	371
Years of incarceration	**Hours of community service**
198	2,860

Source EPA Summary of Criminal Prosecutions Database

Monetary penalty totals are greatly affected by a set of large penalty cases, particularly the prosecution of Equilon Pipeline for a ruptured pipeline, Honeywell International for mislabeling cylinders with harmful chemicals, the explosion at BP's Texas City Refinery, the illegal storage of mercury by the Southern Union Company, and most importantly the prosecution of Transocean for the Deepwater Horizon disaster are notable. The latter prosecution resulted in a $400 million criminal penalty, with all five prosecutions resulting in $501 million in fines or 78% of monetary penalties in our analysis.

Incarceration totals are also influenced by a few significant prosecutions. The top five prosecutions that resulted in the largest amount of incarceration levied at sentencing, including the prosecution of Alan Elias, David Eugene Turner, Arnoldo Carrillo Bazan, Martin Kimber, and Charles Ferris Callihan, resulting in 119 years of incarceration assessed at sentencing, representing 60% of total incarceration. The Turner, Bazan, and Kimber cases might also be thought of as less-traditional environmental crimes (i.e., selling illegal drugs and operating a forced labor camp, a revenge poisoning at a salsa bar, and another revenge-oriented crime centered on a payment dispute). The prosecution of the Orkin Exterminating Company for misuse of a pesticide that killed a couple, whose home was being fumigated, resulted in 2,000 hours of community service or 70% of community service totals in our analysis. The top five probation cases resulted in 71 year of probation or only about 19 of total probation, make these trends less-affected by large penalty cases.

We now turn to describe total restitution awarded to human victims in our data. In cases where restitution is mentioned, is not always clear how much restitution was awarded in the case and if it is specifically awarded to human victims. In some cases, restitution is awarded to EPA or other government agencies to primarily cover the cost of environmental cleanup or private companies and in other cases, the case summaries note that restitution was awarded, but was unclear about the final amount. We find 37 prosecutions or 32% of total prosecutions, involved restitution awarded at sentencing. Of those 37 prosecutions, in five cases it was unclear how much restitution was provided to victims at sentencing. Total restitution awarded to all entities in our estimation was a little over $63 million in the analysis.

In Table 6.2, we describe restitution paid to victims in our analysis, in cases whether the summaries were clear it was going to human victims, their families, or estates, and the amount was clear. We find that over

$25 million was awarded to human victims at sentencing across 21 prosecutions in our data. We approach these figures with caution, given the limitations of a number of cases where restitution is unclear. In the upper-left quadrant, we show that 12 hazardous substances crimes resulted in definable restitution to victims, totaling in excess of $9.24 awarded at sentencing. In the upper-right quadrant, we find that two water pollution prosecutions resulted in $21,234 in restitution to victims that was awarded at sentencing. In the lower-left quadrant, we show that three air pollution crimes were prosecuted, resulting in over $7.75 million in restitution paid to victims. In the lower-right quadrant, we show that four state environmental prosecutions resulted in over $8.3 million in restitution paid to victims.

Notable cases of restitution are rare in our data. A few worthy of mention include: the aforementioned cases of Alan Elias, who exposed employees to cyanide sludge, was ordered to pay $6 million in restitution; Albert David Hajduk, ordered to pay $350,000 to a city POTW worker, who was injured after Luxury Wheels, Inc. illegally discharged plating waste into the sewer; Honeywell International, ordered to pay $2 million in restitution to their victim's family after an exploding cylinder killed a worker; Chemical and Metal Industries were ordered to pay the same amount of restitution to the family in a separate case, having received the cylinder from Honeywell and improperly labeling it before shipping it back to their facility; Arnoldo Carrillo Bazan and his wife, Yini De La Torre, ordered to pay over $974,000 in restitution to poisoning victims; C&J Well Services was sentenced to pay $1.6 million in restitution to a worker's estate after he was killed in an explosion on the job; Green Compass Environmental and the Santa Clara Waste Water Company were each ordered to pay over $2.6 million in restitution to

Table 6.2 Percentage of cases where restitution was awarded to environmental crime victims

Hazardous substances crimes	**Water pollution crimes**
12 Prosecutions	2 Prosecutions
>$9 Million to victims	$21,234 to victims
Air pollution crimes	**State environmental crimes**
3 Prosecutions	4 Prosecutions
>$7 Million to victims	>$8 Million to victims

Source EPA Summary of Criminal Prosecutions Database

victims; and an explosion at an Ashland Oil refinery resulted in $3.75 million in restitution.[6]

Possible Victims

Our case selection for the content analysis in this manuscript resulted in a broad analysis of 117 prosecutions, where we could identify victims of environmental crime from the EPA-CID case summaries. After we completed the content analysis and during the writing of the manuscript, we went back to the prosecution summaries and attempted to identify cases of likely victimization that did not meet our stricter criteria that acute victimization be mentioned in the case summary. We did not include these cases in the analysis throughout the book, but felt it important to draw attention to other potential cases and categories of victimization that were not as acute as injury, death, or evacuation. Our exploration of the data revealed 20 cases that bear honorable mention and give some insight into potential categories of environmental crime victims not acknowledged in our analysis.

In Fig. 6.6, we describe the most common offenses against potential environmental crime victims in our analysis. Across these 20 prosecutions, asbestos crimes make up the most frequent category, where 14 prosecutions center on asbestos violations. In all of these cases individuals were exposed to asbestos because proper protocols were not followed to protect workers and the public from inhaling dangerous asbestos fibers. If we included these cases in our previous analysis, asbestos crimes would represent one of the more common forms of victimization in our data. We chose not to include these cases, as it was not clear that acute injuries or death or evacuation occurred or any other acute victimization in the case. It was also impossible to discern how much asbestos workers and the public were exposed to in the cases. We discuss these 14 cases in kind below to give context to the prosecutions.

Roger Stiles, general manager of Great Pacific Seafoods, Inc. was prosecuted for exposing five workers to asbestos without proper training or protection, including the use of hammers, crowbars, and electric saws to remove asbestos-containing floor tiles.[7] The company was sentenced to serve 60 months of probation and pay $7,000 in restitution to the employees, while Stiles was sentenced to serve 36 months of probation, serve 120 hours of community service, and pay a $5,000 federal fine. Jerry Meide, owner of Meide and Sons ordered employees to remove asbestos

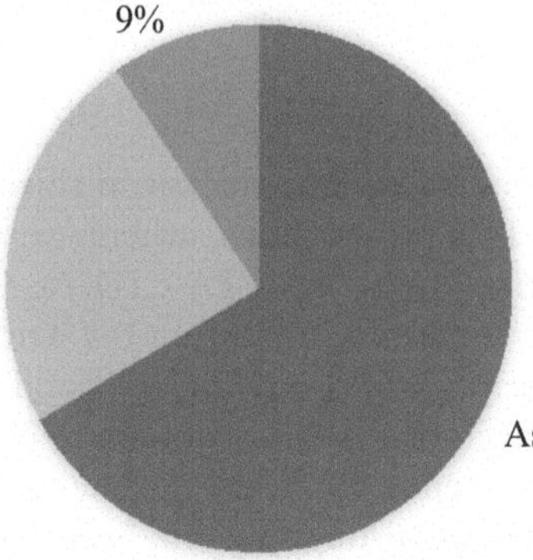

Fig. 6.6 Common offenses against potential environmental crime victims (*Source* EPA Summary of Criminal Prosecutions Database)

pipe insulation from several buildings in Wahpeton, North Dakota without training, protective equipment, and he ordered them to hide the waste material from authorities.[8] Charged with conspiracy and violations of the CAA, the company was ordered to serve 36 months of probation and fined $100,000. Jerry Meide was ordered to serve 10 months of incarceration, 12 months of probation, and to pay $100,000 in restitution to the State of North Dakota.

David Eugene Backus was prosecuted for illegally removing asbestos from a high school in Colorado.[9] The school was contaminated and the facility evacuated, causing $5 million in cleanup costs. The defendant was charged with mail fraud and sentenced to serve 18 months of incarceration, 36 months of probation, and ordered to pay $15,800 in restitution to the school district. Mush Inn and Yoo Jin Management oversaw the removal of asbestos in a boiler room by an unlicensed contractor.[10] Workers were exposed to asbestos and the defendants failed to notify EPA of the asbestos removal. The defendants were convicted for violations of the CAA and were ordered to complete three years of probation and paying both a $35,000 fine and $48,054.09 in restitution to pay for medical monitoring costs for the victims. Co-defendant Tae Ryung Yoon was ordered to serve two years of probation and perform 100 hours of community service.

Roy Charles Bradley was convicted of violations of the CAA for illegally removing asbestos during the demolition of a church, where workers were exposed to asbestos. Bradley was ordered to pay $14,600 in restitution for medical monitoring for four victims, to serve 24 months of probation, and 60 months of incarceration.[11] John Williams, senior corporate manager for Willmax Capital Management and his employer were charged after exposing some 100 people to asbestos during a renovation project in a residential complex in Denver, Colorado.[12] Williams was ordered to serve eight months in prison, serve one year of supervised release, and pay a $100,000 fine. The company was ordered to serve five years of probation and pay for medical monitoring costs for the victims in the case. Grede, LLC was prosecuted after ordering employees to remove asbestos-containing material for an inactive industrial oven without proper training or protection.[13] The company was sentenced for violations of the CAA and was ordered to pay a $200,000 fine and to fund $340,000 in medical monitoring for the 11 affected workers. Joseph Chernis was prosecuted for hiring untrained individuals to remove asbestos pipe insulation.[14] For violations of the CAA, the defendant was ordered to serve 37 months of incarceration and serve three years of supervised release.

Kenneth Baez-Alers, project manager for AIREKO Construction, was prosecuted for removing asbestos from an office building, throwing it in the trash, and failing to notify officials of the removal and release of asbestos-containing materials.[15] For CAA and CERCLA violations, the defendant was ordered to serve 12 months of probation. Philip Farley was prosecuted after directing untrained and unprotected workers to remove asbestos from a 480-unit apartment complex he purchased in St. Petersburg, Florida and throw it in a dumpster.[16] For violations of the CAA, the defendant was ordered to serve a four-year term of probation, three months of home confinement, pay a $8,000 fine, perform 300 hours of community service, and provide $250,000 in restitution to the University of South Florida's College of Public Health's asbestos exposure surveillance fund. Prosecutors charged James Marshall after he lied to workers that testing had shown the cleanout of a building had no asbestos and they continued to work without protection.[17] For violations of the CAA, Marshall was sentenced to serve 12 months of probation and pay $6,332 in restitution to Finger Lakes Occupational Health Services.

Prosecutors indicted Madeline Alonge and four co-defendants for exposing workers to asbestos, failing to conduct air monitoring, and falsifying official documents.[18] The defendants were ordered to pay

$399,000 in restitution to EPA for cleanup costs. Alonge and two co-defendants were ordered to serve three years of probation and two co-defendants were sentenced to eight and ten months respectively and three years of supervised release. Officials indicated Wasatch Railroad Contractors and its CEO, John E. Rimmasch, for exposing 30 of his employees to asbestos during renovation of a historic railcar owned by the National Park Service.[19] For violations of the CAA and obstructing justice, Rimmasch was sentenced to serve 30 months of incarceration, perform 100 hours of community service, serve three years of supervised release, and pay $7,428 in restitution jointly with the company. Prosecutors charged Evans Development Corporation for hiring indigent individuals to remove asbestos without training or protection.[20] The company and the site foreman, Stephen Gassman, were charged with violations of the CAA. The company was ordered to pay a $100,000 fine and $20,000 in restitution. Gassman was ordered to serve six months of home confinement and two years of probation.

Five prosecutions revolved around pesticide crimes. The case profiles mirror those of the previous victim cases we discussed in our analysis. Four involve unpermitted or off-label use of pesticides and one an accident involving pesticides. The cases include the prosecution of Golden West Nuts, Timothy Terrell Smither, North America Green, and two related cases against the Monsanto company.

Prosecutors charged Golden West Nuts after two employees were exposed to pesticides after a hose broke on their equipment and they were told to return to work.[21] The company was charged with state violations and sentenced to serve 35 months of probation, a $20 special assessment fee, and a $10,000 state fine. Co-defendant John Becerra was sentenced to serve 36 months of probation, to pay a $20 assessment fee, and a $100 state fine. Timothy Terrell Smith and his wife Denise Lamber Smither, were prosecuted for conspiracy to commit mail and wire fraud, and violations of FIFRA.[22] The company shipped significant amounts of Termidor, a registered pesticide, for use in indoor taxidermy, although it is not approved for that purpose, and falsified labels, exposing a number of individuals to the pesticide. Timothy Smither was sentenced to serve 12 months and a day in prison, serve 36 months of supervised release, and to pay a $100 special assessment. Denise Smither was sentenced to serve 24 months of probation, with five months of home confinement, and a $25 assessment. Restitution in the case totaling $99,278 was ordered to be paid to their victims.

Prosecutors indicted North American Green after they were found to use a highly toxic chemical, Meth-O-Gas on soil erosion mats.[23] Evidence showed that employees were not properly protected when fumigating and the company violated a number of laws governing the use of the material. For violations of FIFRA, the company was sentenced to serve 36 months of probation, and to pay a $910,000 federal fine. In two cases government officials brought charges against Monsanto for pesticide violations.[24] In the first prosecution, the company was charged with FIFRA violations for spraying a banned pesticide, despite official notice from EPA that it was prohibited. Workers were allowed to enter the treated fields seven days after spraying, when they should have been prohibited for at least 30 days. For FIFRA violations and illegal storage of hazardous waste crimes in violation of RCRA, the company was ordered to pay $10.2 million in fines and payments. In the second prosecution of the company, prosecutors charged Monsanto for using another pesticide and allowing workers to enter a sprayed field well before the time it was safe to do so. The company was ordered to pay $12 million in fines and community service payments.[25]

In two prosecutions the primary crime revolved around hazardous substances. Officials charged Wiley Sanders Truck Lines for illegally transporting more than 64 tons of plastic battery chips.[26] The company did not properly contain the leaking materials, resulting in lead-contaminated residue leaking in the trailers and on public roads. For illegal transportation of hazardous materials in violation of the HMTA, the company was ordered to pay $3 million in fines and community service payments. Prosecutors indicted U.S. Minerals after officials learned employees were exposed to inhalation hazards from the silicate abrasive they manufactured out of waste copper slag from the Anaconda Superfund site.[27] Employees tested high for arsenic and lead in March 2019, despite the fact state officials in Montana shut the facility down in February of that year. For violations of the CAA, the defendant was sentenced to serve five years of probation and pay a $393,200 fine. The company was also ordered to enact a medical monitoring plan for employees at their Montana plant and its other five plants.[28]

Environmental Crimes Resulting in Economic Harm

We uncovered a set of unique cases that did not meet our qualifications for victimization in the broader analysis but bear some mention here for context into victimization in environmental crime prosecutions. In these cases, the defendants committed environmental crimes that caused economic damages to defendants, although they may not have injured them directly. These include the prosecution of Robert J. Lucas, Wally K El-Beck, David Fredericy, and the Scotts Miracle-Grow Company.

Robert J. Lucas was prosecuted as part of an illegal development that engaged in the unauthorized filling of wetlands without a permit to create a 2,600 acre subdivision in Vancleave, Mississippi, known as Bill Hill Acres.[29] While Lucas and his five co-defendants did not injure, kill, or cause the public to be evacuated, they committed an extensive environmental crime that defrauded more than 600 families left with raw sewage seeping out of the ground, failed septic systems, sinkholes, and foul-smelling backup in homes. For conspiracy and violations of the CWA, the two companies were sentenced to cumulatively serve 120 months of probation, and to pay over $5.8 million in monetary penalties. Lucas and the remaining co-defendants were cumulatively sentenced to pay over $905,000 in monetary penalties, serve 108 months of probation, and 282 months of incarceration.[30] Prosecutors charged Wally K. El-Beck for falsely accepted hazardous waste for incineration and issuing Certificates of Destruction of the waste to his customers.[31] We have found by investigators to have illegally stored some 10,000, 55-gallon drums of industrial waste at his company's site. For defrauding clients out of $1.16 million, the defendant was charged with 58 counts of mail fraud and was sentenced to 18 months of incarceration, 24 months of probation, to pay a $3,800 special assessment fee, and restitution over $904,961.10.

Prosecutors indicated David Fredericy and Joseph Kuzlik for conspiring to violate civil rights, when they contaminated a residential property with mercury in Cleveland, Ohio, because of their race and making false statements to officials.[32] Fredericy was sentenced to 33 months of incarceration, 36 months of probation, a $300 special assessment fee, and to pay $26,701 in restitution to the Ohio EPA. Kuzlik was sentenced to serve 27 months of incarceration, 36 months of probation, and pay a $300 special assessment fee, as well as $26,701 in restitution to the Ohio EPA and his victims. Officials charged the Scotts

Miracle-Grow Company after it applied pesticides to bird food, although EPA had prohibited such use.[33] Scotts sold 70 million units of bird food contaminated with a pesticide that kills birds and fish, before voluntarily recalling the product. The company also illegally sold the pesticides and marketed them with misleading claims. The company was ordered to pay a $4 million federal fine and make $500,000 in payments.[34]

Conclusion

In this chapter, we worked to develop a broader victimology of environmental crime victims that built on the analysis from the three previous chapters. Our framework examined who was most likely to be victimized, the consequences of their victimization, the most common offender behaviors, punishment patterns for offenders, and exploration of cases that did not fit the criteria for inclusion in our analysis, but still whose mention has something important to say about the role of victims in environmental crime prosecutions. We take some time here to discuss a few of the key implications of the chapter, before turning to the final chapter, which discusses these findings in greater detail and discusses prospects for enhancing the protection of environmental crime victims within the legal system.

We find in this chapter that workers are the most common environmental crime victims in environmental crime prosecutions. Thirty-six percent of prosecutions involve individuals injured at work. In 31% of cases, one or more individuals were victimized by companies. In 23% of cases, one or more individuals were victimized by another individual offender. Thus, when discussing victimization in environmental crime prosecutions, our analysis suggests companies are the most common offender, first with people injured on the job, then the public, and finally individuals injuring others.

A second takeaway from the analysis is the categorization of offenses that lead to victimization. We find that hazardous substances crimes make up half of all cases in the analysis and constitute the most common category of victimization. The second most common category is water pollution crimes, making up 19% of cases. Air pollution crimes are responsible for 16% of prosecutions and state environmental crimes constitute 15% of the cases in our analysis. These patterns follow broader trends in charging patterns, where violations of the RCRA,CWA, and the CAA are the most common violations in our data.

When we parcel out the types of offending behaviors from the categories of crime, about 59% can be organized into three categories. The most common act that lead to victimization was an explosion, constituting 31% of cases in our analysis. The second most common offense was a pesticide crime, leading to 17% of the prosecutions in our data. Lead-based paint crimes were the third most common offender behavior, resulting in 11% of prosecutions in the analysis. Summarily, from these findings we can state that negligent and knowing actions by companies resulting in explosions, particularly those that kill or injure workers, tend to be one of the most common offender behaviors we see across these prosecutions, followed by pesticide crimes.

The consequences of victimization can be severe in our data. Death or a combination of death and injury occur in 25% of the cases in our data. Evacuations or a combination of evacuations and injuries constitute 21% of the prosecutions in our data. Injuries were by far the most common consequence of offenders' behaviors in our analysis.

The consequences of offending could also be severe, but stiff penalties are generally uncommon. While some $610 million in monetary penalties were assessed to defendants at sentencing in our analysis, alongside 371 years of probation, 198 years of incarceration, and 2,860 hours of community service, a handful of significant prosecutions greatly affect these overall totals, particularly in all cases but probation. While these may have some deterrent value within the broader universe of potential individuals and companies that may have otherwise victimized the public or people at work, this is difficult to know. To the best of our ability, we estimated over $25 million of these penalties were in the form of restitution to human victims, their families, or estates.

Notes and References

1. Kenneth Douglas Griffin (California 2015023881, 2017).
2. Airosol Company, Inc. (D. Kansas 6:21CR10074, 2022).
3. Cecil Morgan (W.D. Louisiana CR 94-20061-02, 1995).
4. David Lynn Jensen (Utah 021908862FS, 2003).
5. Steve Eiffler (Utah 051903875 FS, 2006).
6. In two significant cases we made judgement calls on the final data entered for the analysis. In the first, the Alan Elias case, he was ordered to pay "almost $6 million in restitution" and we choose to note it in the analysis as $6 million. In the second case against

Ashland Oil, $3.75 million in restitution was paid to "the five employees injured in the explosion and to several fire departments" and we chose to keep this restitution to victims here.

7. Great Pacific Seafoods, Inc. (D. Alaska A01-015, 2001).
8. Meide and Son, Inc. (D. North Dakota C3-98-68, 1999).
9. David Eugene Backus (D. Colorado 01-CR-103N, 2005).
10. Mush Inn (D. Alaska 3:19-CR-00107, 2021).
11. Roy Charles Bradley (E.D. Michigan 1:13-CR-20622, 2017).
12. John Williams (D. Colorado 16-CR-00355-LTB, 2017).
13. Grede, LLC (E.D. Wisconsin 17-CR-50, 21018).
14. Joseph Chernis (C.D. Illinois CR16-30033, 2018).
15. Kenneth Baez-Alers (D. Puerto Rico 3:15CR871, 2018).
16. Philip Farley (M.D. Florida 8:15-CR-00133-RAL-MAP, 2018).
17. James Marshall (W.D. New York 6:20-CR-06003, 2021).
18. Madeline Alonge (N.D. New York 1:21-CR-00136, 2022).
19. Wasatch Railroad Contractors (D. Wyoming 21-CR-00138,2022).
20. Evans Development Corporation (W.D. Kentucky 3:99CR-107-01-R, 2000).
21. Golden West Nuts, Inc. (California CAL 2004-401, 2007).
22. Timothy Terrell Smither (E.D. North Carolina 5:11-CR-371-1FL, 2012).
23. North American Green (S.D. Indiana 3:11-CR-0050RLY-WGH, 2012).
24. Monsanto (D. Hawaii 1:19CR00162, 2020).
25. Monsanto (D. Hawaii 1:21CR00140, 2022).
26. Wiley Sanders Truck Lines (C.D. California CR 19-00035-PA, 2019).

    United States Attorneys' Office, 2019, Trucking Company Ordered to Pay $3 Million for Illegally Transporting Hazardous Materials from the Exide Battery Recycling Plant in Vernon, Available from: https://www.justice.gov/usao-cdca/pr/trucking-company-ordered-pay-3-million-illegally-transporting-hazardous-materials-exide.
27. U.S. Minerals, Inc. (D. Montana CR-21-25-BU-DLC, 2022).
28. United Stats Attorney's Office, 2019, U.S. Minerals, Inc. Sentenced for Clean Air Act Violation that Exposed Employees to Arsenic at former Anaconda Plant, Available from: https://www.justice.gov/usao-mt/pr/us-minerals-inc-sentenced-clean-air-act-violation-exposed-employees-arsenic-former.

29. Robert J. Lucas, Jr. (S.D. Mississippi 1:04CR60GuRo-001, 2006).
30. U.S. Department of Justice, 2005, Defendants Receive Major Jail Sentences, Pay Restitution for Major Wetlands Criminal Prosecution, Available from: https://www.justice.gov/archive/opa/pr/2005/December/05_enrd_644%20.html.
31. Wally K. El-Beck (E.D. Arkansas 4:05CR00179GTE, 2008).
32. David Fredericy (N.D. Ohio 1:06CR035, 2007).
33. Scotts Miracle-Grow Company (S.D. Ohio 2:12-CR-00024-JLG, 2012).
34. We assumed in this case the company killed more than its fair share of its customers' birds and thus under the law, damaged their property.

CHAPTER 7

Protecting Victims of Environmental Crime

Abstract Our analysis has moved toward developing a better understanding of environmental crime victims in the United States, through an analysis of environmental crime prosecutions. Examining almost four decades of EPA criminal investigations that lead to prosecution has allowed us to generate significant insights into who is most likely to be victimized, by whom, where, the categories of victimization, and the punishment and restitution patterns meted out to offenders at sentencing. In this chapter, we return to some of our key takeaways from the analysis, followed by a discussion of trends in prosecution over time and the possibilities of criminal deterrence, and then discuss obstacles to enhanced criminal enforcement efforts and practical possibilities for deterring offenders and better protecting victims in the legal system in the United States.

Keywords Protecting victims · Deterrence · Stronger enforcement

J. Ozymy and M. Jarrell Ozymy, *The U.S. Administrative State and the Protection of Environmental Crime Victims*, Palgrave Studies in Green Criminology, https://doi.org/10.1007/978-3-031-62680-7_7

Key Findings

The most likely scenario of environmental victimization we find in our analysis occurs at work. Thirty-six percent of our cases involve employees injured on the job. Alternatively, companies and their officers are the most common offenders, where 67% of prosecutions involve victimization at work or companies victimizing the public. In 23% of prosecutions, offenders are individuals that victimize others.

When we examine the causes of victimization, the most prevalent category that presents is an explosion, which was the driver of the victimization in 31% of the cases in our data. After explosions, offenses involving pesticide crimes (17% of prosecutions) were the second most common way someone could be victimized, followed by exposure to lead-based paint, which occurred in 11% of prosecutions. The remaining 41% of prosecutions are broadly construed. When we considered cases of "possible victimization" in the previous chapter, we found that asbestos crimes were the most prevalent (67% of cases), followed by pesticide crimes (24%), and hazardous substances crimes (9%).

When we typologize the outcomes of victimization, being injured in the commission of an environmental crime was by far the most common situation, occurring in 53% of cases. Death or the combination of death and injury was the second most common outcome in our analysis at 25% of prosecutions under consideration. Being evacuated and injured or simply evacuated and removed from your place of residence because of an environmental crime was the third most common outcome of an environmental offense, constituting about 21% of prosecutions.

If explosions were the most common, singular reason people were victims of environmental crime in our analysis and individuals were predominantly injured on the job, exposure to hazardous substances represented the most common category of offense in our analysis. Roughly half of all crimes under consideration in our analysis result from exposure to hazardous substances, such as chemicals, lead-based, paint, and pesticides. Being the victim of a water pollution crime, constituting 19% of prosecutions, was the second most common category of offense we found. Being a victim of an air pollution crime, representing 16% of prosecutions, was the third most common offense, and being victimized during the commission of a state environmental crime rounds out our categories of victimization at 15% of prosecutions under consideration in the analysis.

Finally, the consequences for offenders could be severe, but such punishments were uncommon in our analysis. Offenders were collectively sentenced to pay over $640 million in monetary penalties, to serve 371 years of probation and 198 years of incarceration, and to perform 2,860 hours of community service. Monetary penalties, incarceration, and community service, were found to be significantly impacted by a handful of outlier prosecutions, whereas probation totals were less skewed toward these significant prosecutions. Restitution was more difficult to estimate, given that numbers were often unknown at the time of sentencing, but to the best of our ability we would suggest at least $25 million was assessed to offenders at sentencing and was specifically directed toward human victims, their families, or their estates. These cases constitute 21 prosecutions or 18% of the cases in our analysis.

Trends in Prosecutions and Prospects for Deterrence

Taking a broad view environmental victimization requires exploring prospects for deterrence in the current regulatory and legal regimes and whether we might speculate of a future that better protects victims. We found in our analysis that prosecutions with identifiable human victims were rare and often concentrated in particular areas of the country, which on its face perhaps does not bode well for conceptualizing criminal enforcement as a strong deterrence mechanism for offenders. We return to some of these trends with additional analysis in Fig. 7.1, where we describe the number of prosecutions with identifiable human victims by EPA fiscal year, 1983.

When we examine prosecutions holistically, rather than parceled by the categories in the previous chapters, prosecutions are still infrequent over time. The first prosecutions in our data area adjudicated in 1989 and 1990, which is unsurprising, given the criminal enforcement apparatus was only beginning to institutionalize in the mid-1980s. From 1990 to 1999, we find prosecutions increase, totaling 18 during the decade and reaching an annual high point of six prosecutions adjudicated in 1998. From 2000 to 2009 prosecutions continue to increase, with 30 cases adjudicated during the decade and an annual high of six prosecutions in 2003 and 2007, respectively. The next decade shows another increase in prosecutions, rising to 55 prosecutions from 2010 to 2019, with an annual high of eight prosecutions adjudicated in 2012, 2017, and 2018.

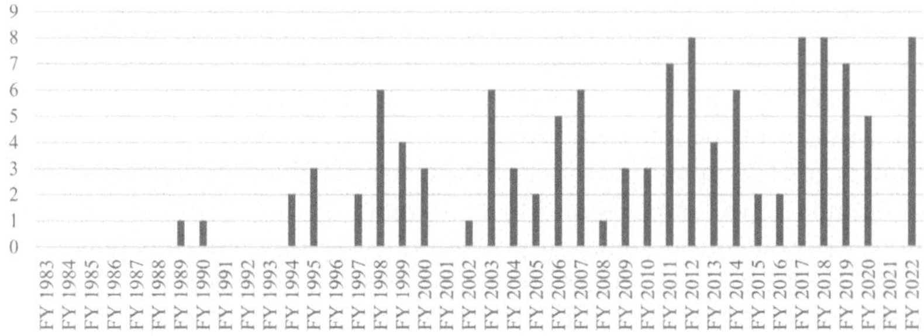

Fig. 7.1 Number of prosecutions with environmental crime victims by EPA Fiscal Year, 1983–2022 (*Source* EPA Summary of Criminal Prosecutions Database)

Thirteen prosecutions were adjudicated from 2020 to 2022, making a grand total of 117 cases adjudicated, with an annual average of about 2.9 prosecutions.

While prosecutions with identifiable human victims have been rare since the early 1980s when the criminal enforcement regime institutionalized, we do see an uptick over time. Whether this presents a deterrent effect is hard to know. In the past two decades, 88 prosecutions were adjudicated, averaging about 4.6 a year, while much better than the historical averages since 1983, are still fairly low. When we look at the geography of these prosecutions and victims, we find they are also heavily concentrated in a few states. We describe these trends in Fig. 7.2, which shows total prosecutions adjudicated per U.S. state and territory since 1983.

Prosecutions are spread across 41 states, Puerto Rico, and the U.S. Virgin Islands. Most of the states and territories have only a few prosecutions over time, representing no more than 2–3% of total cases. With 15 prosecutions adjudicated or 13% of cases in our analysis, California is the leader in the analysis, followed by Louisiana (9% of prosecutions), New York (6% of prosecutions), and Texas (6% of prosecutions). These four states make up 34% of prosecutions in our data. The limited numbers outside of a few states also cast some doubt on the prevalence of prosecutions that may be required to deter offenders, but this is a complex phenomenon, given the broad nature of the federal and state civil and criminal enforcement regimes.

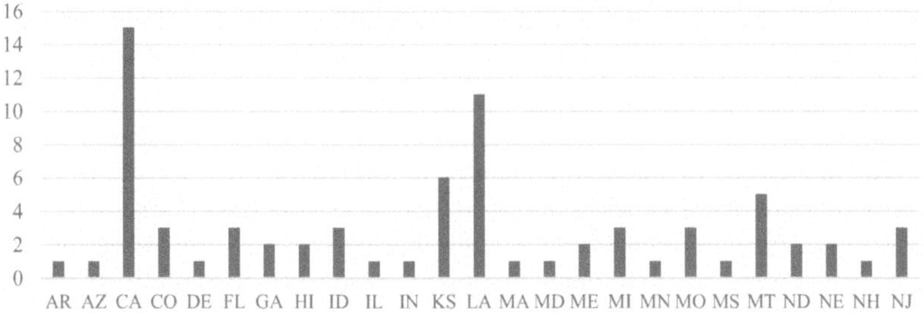

Fig. 7.2 Number of prosecutions with environmental crime victims by U.S. State and Territory, 1983–2022 (*Source* EPA Summary of Criminal Prosecutions Database)

Obstacles to Stronger Criminal Enforcement

We can suggest that the most obvious path toward reducing environmental victimization that flows from our analysis, is to better protect workers on the job. When corporate officers, owners, and other company officials engage in negligent or knowing behaviors they hurt people, but particularly their employees. A safer workplace, particularly when industries deal with toxic, flammable, and explosive materials would go a long way toward reducing victimization. In fairness, companies also hurt the public as well in a number of ways illustrated herein. Releasing toxic chemicals into the ambient air, spilling them on the ground, or illegally discharging them into POTWs are common pathways that companies victimized the public in our data. Outside of taking stronger safety measures to protect workers, companies should also take simple steps, such as following proper safety and labeling protocols for registered pesticides—a significant source of unnecessary victimization that seemed in almost all cases to be avoidable—and following straightforward lead-based paint regulations and disclosures, specifically targeted at children, who are the likely victims of ingesting toxic lead from paint chips or employees exposed to lead in the workplace. From the best we could glean, in less than one in five prosecutions, restitution was ordered to help make victims and their families whole, or medical monitoring was ordered in some cases at the offender's expense. Looking forward, a more successful criminal enforcement regime would enhance ways to prioritize cases that involve victims, deter offenders from victimizing their workers and the general public, and take a restorative justice approach to victims and their communities when they are damaged from an environmental crime. Obstacles to enhanced enforcement on behalf of victims are numerous and structural in nature.

The first major obstacle to enhanced enforcement is the nature of environmental law itself in the United States. Congress has never chosen to pass an overarching environmental law that crosses environmental media, yet many environmental problems cross various media, limiting EPA's regulatory scope to specific issues, rather than being able to pass broader regulations that cross boundaries for the sake of improved public and environmental health.[1] A related problem is that federal environmental law has been "in arrears" for a very long time.[2] Major amendments to RCRA, CWA, CAA, and others that have not occurred for over three decades are arguably in need of modernization.[3] This would appear most obviously with the inability of the EPA to regulate carbon emissions under the CAA, as evidenced in the recent Supreme Court decision in *West Virginia v. EPA* that limited the scope of President Obama's Clean Power Plan to allow EPA to regulate carbon emissions, specifically those generated by coal-fired power plants.[4] The court focused on issues pertaining to the "major questions doctrine" about what EPA is allowed to regulate outside of the scope of what Congress was specifically clear about when writing the CAA.[5] While this is the most salient example of the limits of EPA policymaking in a world of moribund federal environmental law, the principle applies to most major questions in other statutes, such as defining the boundaries of navigable waters under the CWA, or forcing needed risk assessments for a variety of pesticides under FIFRA, and the list continues extensively.[6]

The regulatory structure in the United States environmental scheme is also very decentralized, with state agencies taking up the bulk of the civil and administrative enforcement actions. It is hard to know the extent of their criminal enforcement actions over time, as no consistent data source has yet been assembled to tackle such an empirical puzzle, but we suspect, particularly given the complexity and resources needed to investigate environmental crimes, particularly those committed by well-resourced corporations, that the states have not taken the lead on this role historically. Our analysis gives us some limited sense of these efforts, at least in terms of those investigations where EPA-CID collaborated with state agencies and cases were differed for state-level prosecutions. We found that 15% of victim-focused prosecutions in our data happened at the state level, but those numbers are limited to 17 prosecutions over almost four decades. While only a limited view into the world of state criminal enforcement efforts for victims, it would not surprise us if it were possible to gather all of the data in the universe, the results would increase much as a percentage of the overall criminal enforcement efforts undertaken by the states and the EPA.

Another major hurdle to enhancing criminal enforcement is resources. For decades, at various levels, both the Democratic and Republican Parties have chosen to underinvest in stronger enforcement for the long haul.[7] The former has chosen to support, but not prioritize enforcement on the whole, and for the most part, the latter has ranged from disinterested to hostile to enforcement.[8] The result of this back and forth like so many regulatory policies in the United States has been a stalemate. The EPA has been "running on fumes" for a number of years, attempting to meet its organizational objectives in this ungenerous environment.[9] Examining the budgetary situation for EPA and ENRD gives the broader strokes. Inflation-adjusted budgetary allotments for EPA reached a high point in the late 1970s, where nominal staffing reached its peak in 1999.[10] For the last decade, ENRD's budget has also been fairly stagnant when adjusted for inflation.[11] Returning to another important metric of underinvestment, the number of EPA-CID criminal investigators mirrors budgetary disinvestment trends. The Pollution Prosecution Act of 1990 raised the hiring floor for EPA-CID criminal investigators to 200, but depending on source, that number has fluctuated at or below that goal since the 1990s.[12] The well of support that percolated up from some bipartisan enthusiasm from stronger penalties for all types of federal crimes in the 1980s waned considerably by the time George W. Bush entered office and has not returned and is unlikely to in the near future.[13] By the time Donald Trump left office, environmental enforcement may have reached a new low in terms of morale and support, with 700 EPA employees leaving or retiring, enforcement actions declined, as did funding.[14]

Prospects for Change

With so many political and structural obstacles to enhancing criminal enforcement efforts to protect victims of environmental crime, prospects for change are possible in the short and long term. While the nature of the political and regulatory system in the United States are structural features that will likely resist major overhaul, the Biden Administration has made overtures to victims indirectly with its investment in protecting environmental justice communities. The FY 2022 federal budget committed $11 billion to EPA and raised staffing numbers to 15,000 and while this does not come near to returning the agency to its high staffing numbers of the late 1990s, it is a positive movement in the right direction, even if the enacted budget was $9.5 billion and 14,581 staff.[15] ENRD's appropriation of $133 million was not historically significant either adjusted for inflation, but still trending in the better direction.[16]

The Biden Administration's Justice 40 Initiative is the single largest commitment to enhancing funding and enforcement in environmental justice communities in U.S. history.[17] Building off earlier efforts that began to institutionalize environmental justice concerns during the Clinton Administration, the Initiative invests federal resources into disadvantages communities to address energy, water infrastructure, climate change, and working to remediate legacy pollution.[18] Executive Order 14,008 also created the White House Environmental Justice Interagency Council to coordinate between the White House and environmental justice organizations.[19] Moreover, the Administration required environmental agencies to refocus their enforcement efforts to ameliorating harm in environmental justice communities, which has become a focus of EPA and DOJ's efforts in the subsequent year.[20] As environmental justice communities are the most burdened communities in the country in terms of environmental harm, the act of focusing resources and enforcement toward their needs and enhancing collaborations, should go much farther than previous, piecemeal efforts toward centering enforcement on environmental crime victims in these communities.

Focusing enforcement efforts on environmental justice communities must be paired with structural investments in the criminal enforcement apparatus that continue this trend of funding set in the Biden Administration. Meeting standards for EPA-CID criminal investigators under the Pollution Prosecution Act and then exceeding them is one place to start. Expecting criminal deterrence with so few staff to monitor and police violations seems unreasonable.[21] The number of specialized prosecutors, currently numbering 43 in DOJ-ECS, could arguably be increased significantly in order to enhance caseloads and prosecutions for deterrence.[22]

Funding enhancements are also needed for state environmental law enforcement agencies that handle the largest share of enforcement efforts.

Enhancing state efforts can free up time for complex prosecutions to be pursued by federal environmental law enforcement agencies. Certainly lawsuits have a place in compliance and enforcement too, as do criminal enforcement efforts, but strong collaboration with the states and a good working relationship is needed to make the system function more effectively.[23] A related solution enhancing relationship building and collaborations is further support for criminal justice associations that help to train investigators and build collaborations.[24]

Criminal enforcement enhancements need not be the only solution to reduce victimization and other strategies can prevail alongside deterrence. For example, strong civil compliance programs can help prevent serious violations of law before they occur.[25] Civil remedies can also include a number of restorative justice elements that focus on returning communities to the condition before they are harmed.[26] These principles may take the form of environmental and health monitoring or compensatory schemes for victims previously discussed in our analysis and these could be extended by the courts to a greater number of situations. Stronger compliance measures, civil enforcement, enhanced resources for state agencies and coordination, will all bolster the overall enforcement scheme and work to foster stronger protection of victims within the administrative state.

A final remedy to better protect environmental crime victims under the law and within the criminal enforcement process is to afford them procedural rights within criminal proceedings. Over the last two decades, we have seen a resurgence of the idea that granting procedural rights to crime victims is warranted within the criminal justice system in the United States.[27] Early concepts of individual rights focused on rights not of individuals, but through kinship ties, clans, or other cultural associations.[28] In Colonial America, it was still somewhat common for victims of crime to play a role in court proceedings, but this role was diminished overtime and replaced with a stronger role for the prosecutor.[29] Well into the 1800s, however, private prosecutions were still common in America.[30] As the process of criminal proceedings became standardized over time, the role of victims significantly diminished, or as the court noted in the famous case of *Kenna v. U.S. District Court*, victims of crime should be seen but not heard, just like Victorian children.[31]

This vision for victims has been changing for some time. In 1965, California passed legislation allowing compensation to be sought for victims of violent crime, which was followed by New York that created the Crime Victims Compensation Board now administrated by the state Office of Victim Service (OVS).[32] President Reagan declared April 19 to be "National Victims' Rights Week" in 1982 and created a taskforce that provided a Final Report on Victims of Crime.[33] Congress passed the federal Victims of Crime Act (VOCA) in 1984.[34] The Office for Victims of Crime (OVC), was established in 1988 and overseen by DOJ, administers the Crime Victims Fund (CVF) to aid and possibly compensate crime victims.[35]

The federal courts have established legal precedent that environmental crime victims should be awarded procedural rights in court. The CVRA defines a victim as "person directly or proximately" harmed in the commission of an offense.[36] In three specific environmental crime prosecutions including, *United States v. BP Products,* where an explosion at British Petroleum's Texas City refinery that occurred on March 23, 2005, which killed 15 workers and injured 180 workers, *United States v. W.R. Grace & Co,* where the company mined vermiculite near Libby Montana from 1960 to 1990 and exposed numerous individuals to asbestos, and *United States v. CITGO,* have all resulted in some additional procedural rights for victims of environmental crime.[37] In additional to regulatory enforcement solutions and criminal prosecution, placing environmental crime victims at the center of prosecutions could be a powerful tool for prosecutors to focus their efforts to combat serious environmental crimes and protect, prioritize, and take a restorative justice approach to victims in the process.[38]

Notes and References

1. R. Arnold and A.B. Whitford, 2005, Organizational Dilemmas of the U.S. EPA: Why Structure Matters for Environmental Protection, 14 Environmental Politics, 118–123.
2. Richard J. Lazarus, 2013, Environmental Law has Fallen 'In Arrears", Harvard Law Today, Available from: https://hls.harvard.edu/today/richard-lazarus-environmental-law-has-fallen-in-arrears-video/.
3. The classic case here is *Chevron U.S.A., Inc. v. Natural Resources Defense Council, Inc.*, 82-1005 (1984). The *Chevron* case established a precedent of potential deference by the Supreme Court to administrative agencies in issues of policymaking. The issue at hand was amendments to the CAA, where Congress did not clearly

define "source" for the purposes of defining whether a company would need to submit to a review process for updating or major additions to a facility or the parameters of that permitting.

4. *West Virginia v. EPA*, 20-1530 (2022).

   John Copeland Nagle, 2017, An Autopsy of the Clean Power Plan, Ecology Law Quarterly Currents, Available from: https://www.ecologylawquarterly.org/currents/an-autopsy-of-the-clean-power-plan-john-copeland-nagle-the-clean-power-plan-cpp-was-supposed-to-be-great-the/.

   Adam Liptak, 2022, Supreme Court Limits E.P.A's Ability to Restrict Power Plant Emissions, New York Times, Available from: https://www.nytimes.com/2022/06/30/us/epa-carbon-emissions-scotus.html.

5. This doctrine was addressed earlier in the important case of *Utility Air Regulatory Group v. Environmental Protection Agency*, 573 U.S. 302 (2014).

   Kevin O. Leske, 2016, Major Questions About the "Major Questions" Doctrine, 5 Michigan Journal of Environmental & Administrative Law, 479–451.

6. Michael J. McClary and Jessica B. Goldstein, 2017, FIFRA at 40: The Need for Felonies for Pesticide Crimes, 47 Environmental Law Reporter, 10767.

7. Joel A. Mintz, 2005, Neither the Best of Times Nor the Worst of Times: EPA Enforcement During the Clinton Administration, 35 Environmental Law Reporter, 10390.

8. Lia Cattaneo, 2021, Wrinkling Citizen Suits: *California V. EPA* (9th Cir 2020) and Clean Air Act Underenforcement, 45 Harvard Environmental Law Review, 503.

   John Peter Suarez, 2003, Management Review of the Office of Criminal Enforcement, Forensics and Training, 5–7, Available from: https://www.epa.gov/sites/production/files/documents/oceft-review03.pdf.

   Cally Carswell, 2017, How Reagan's EPA Chief Paved the Way for Trump's Assault on the Agency, Available from: https://newrepublic.com/article/141471/reagans-epa-chief-paved-way-trumps-assault-agency.

9. Joel A. Mintz, 2016, Running on Fumes: The Development of New EPA Regulations in an Era of Scarcity, 46 Environmental Law Reporter, 10510-19.

10. U.S. EPA, 2024, EPA's Budget and Spending, Available from: https://www.epa.gov/planandbudget/budget.

    U.S. Inflation Calculator, 2024, Available from: https://www.usinflationcalculator.com/.

11. U.S. Department of Justice, 2024, Budget and Performance Summary ENRD (Various Years), Available from: https://www.justice.gov/doj/budget-and-performance.

    U.S. Department of Justice, Environment and Natural Resources Division, 2022, FY 2023 Performance Budget, 15, Available from: https://www.justice.gov/jmd/page/file/1491706/download.

12. The Pollution Prosecution Act of 1990 (P.L. 101-593).

    U.S. EPA, 2024, U.S. Environmental Protection Agency Criminal Enforcement Program: America's Environmental Crime Fighters. Available from:

    https://www.epa.gov/sites/production/files/documents/oceftbrochure.pdf.

    Public Employees for Environmental Responsibility (PEER), 2019. EPA CID Agent Count, Available from: https://www.peer.org/wp-content/uploads/2019/11/11_21_19-Federal_Pollution_EPA_CID_Agent_Count.pdf.

13. Lia Cattaneo, 2021, Wrinkling Citizen Suits: *California V. EPA* (9th Cir 2020) and Clean Air Act Underenforcement, 45 Harvard Environmental Law Review, 503.

    Richard J. Lazarus, 1994, Assimilating Environmental Protection into Legal Rules and the Problem with Environmental Crime, 27 Loyola Law Review, 867–870.

    Judson W. Starr, 1991, Turbulent Times at Justice and EPA: The Origins of Environmental Criminal Prosecutions and the Work that Remain, 59 George Washington Law Review, 900-905.

    Enforcement efforts increased under G.W. Bush, but resources were eventually focused toward the War on Terror.

    David M. Uhlmann, 2008, Strange Bedfellows, 25 Environmental Law Forum, 40–44.

14. Cally Carswell, 2017, How Reagan's EPA Chief Paved the Way for Trump's Assault on the Agency, Available from: https://newrepublic.com/article/141471/reagans-epa-chief-paved-way-trumps-assault-agency.

Seven hundred employees exited the EPA under Trump, dropping the number of staff to 14,172, see: Public Employees for Environmental Responsibility, 2017, 700+ Employees have Left the EPA Under Trump: Loss of Scientists, Staffers Undermines Agency's Purpose, Available from: https://peer.org/700-employees-have-left-the-epa-under-trump-loss-of-scientists-staffers-undermines-agencys-purpose/.

Elgie Holstein, 2017, The Severe, Real-World Casualties of Trump's EPA Budget Cuts, Available from: https://www.edf.org/blog/2017/03/03/severe-real-world-casualties-trumps-epa-budget-cuts.

Valerie Volcovici, 2017, U.S. EPA Employees Protest Trump's Pick to Run Agency, Reuters, Available from: https://www.reuters.com/article/usa-epa-pruitt-idUSL1N1FR1NZ.

15. U.S. EPA, 2024, Statement by Administrator Regan on the President's FY 2022 Budget, Available from: https://www.epa.gov/newsreleases/statement-administrator-regan-presidents-fy-2022-budget.
16. U.S. Department of Justice, Environment and Natural Resources Division, FY 2023 Performance Budget, 15, Available from: https://www.justice.gov/jmd/page/file/1491706/download.
17. U.S. EPA, 2024, EPA Launches New National Office Dedicated to Advancing Environmental Justice and Civil Rights, Available from: https://www.epa.gov/newsreleases/epa-launches-new-national-office-dedicated-advancing-environmental-justice-and-civil.
18. The White House, 2024, Justice40: A Whole-Of-Government Initiative, Available from: https://www.whitehouse.gov/environmentaljustice/justice40/.
19. Executive Order 14,008 also created the White House Environmental Justice Interagency Council (IAC) to create collaborations between the White House and local environmental justice groups to address historical injustices, see: The White House, 2022, White House Interagency Council, Available from: https://www.whitehouse.gov/environmentaljustice/white-house-environmental-justice-interagency-council/.
20. U.S. EPA, 2024, New Enforcement Strategy Advances President Biden's Environmental Justice Agenda, Available from: https://www.epa.gov/newsreleases/new-enforcement-strategy-advances-president-bidens-environmental-justice-agenda.

U.S. Department of Justice, 2021, Environment and Natural Resources Division Distributes Memorandum Summarizing Enforcement Policies and Priorities, Available from:

https://www.justice.gov/opa/pr/environment-and-natural-resources-division-distributes-memorandum-summarizing-enforcement.

U.S. EPA, 2024, Environmental Justice in Enforcement and Compliance Assurance, Available from: https://www.epa.gov/enforcement/environmental-justice-enforcement-and-compliance-assurance.

21. Carole M. Billiet and Sandra Rousseau, 2014, How Real Is the Threat of Imprisonment for Environmental Crime? 37 European Journal of Law and Economics, 183–188.

    Michael J. Lynch, 2017, The Sentencing/Punishment of Federal Environmental/Green Offenders, 2000–2013, 38 Deviant Behavior, 991–995.

22. DOJ-ECS, 2021, Prosecutors Protecting Our Nation's Ecological Heritage, Available from: https://www.justice.gov/enrd/environmental-crimes-section.

23. Stephen M. Johnson, 2014, Sue and Settle: Demonizing the Environmental Citizen Suit, 37 Seattle University Law Review, 891.

24. Henry N. Butler, Nathaniel J. Harris, 2014, Sue, Settle, and Shut Out the States: Destroying the Environmental Benefits of Cooperative Federalism, 37 Harvard Journal of Law and Public Policy, 579.

    Raymond W. Mushal, 2009, Up From the Sewers: A Perspective on the Evolution of the Federal Environmental Crimes Program, Utah Law Review, 1103–1107.

25. Lucia Ann Silecchia, 2011, Ounces of Prevention and Pounds of Cure: Developing Sound Policies for Environmental Compliance Programs, 7 Fordham Environmental Law Journal, 583.

26. Michael L. Rustad, Thomas H. Koenig, and Erica R. Ferreira, 2013, Restorative Justice to Supplement Deterrence-Based Punishment: An Empirical Study and Theoretical Reconceptualization of the EPA's Power Plant Enforcement Initiative, 2000–2011, 65 Oklahoma Law Review, 427.

27. Paul G. Cassell, 2005, Recognizing Victims in the Federal Rules of Criminal Procedure: Amendments in Light of the Crime Victims' Rights Act, Brigham Young University Law Review, 835.

    Paul G. Cassell, 2015, Introduction: The Maturing Victims' Rights Movement, 13 Ohio State Journal of Criminal Law, 1–2.

    Paul G. Cassell, 2007, Treating Crime Victims Fairly: Integrating Victims into the Federal Rules of Criminal Procedure, Utah Law Review, 861.
28. Stephen Shafer, 1992, Victimology: The Victim and His Criminal (Reston).
29. William F. McDonald, 1976, Towards a Bicentennial Revolution in Criminal Justice: The Return of the Victim, 13 American Criminal Law Review, 649.
30. Douglas E. Beloof, 2007, Weighing Crime Victims' Interests in Judicially Crafted Criminal Procedure, 56 Catholic University Law Review, 1135.
31. Mario M. Cuomo, 1992, The Crime Victim in a System of Criminal Justice, 8 Journal of Civil Rights and Economic Development, 3.

    Kenna v. U.S. District Court: 435 F.3d 1011 (2006).

    In *Kenna,* the court made the case for a right for victims to be heard, by making statements at sentencing. See: National Crime Victim Law Institute, 2011, Fundamentals of Victims' Rights: A Summary of 12 Common Victims, 11, Available from: https://law.lclark.edu/live/files/11823-fundamentals-of-victims-rights-a-summary-of-12.
32. Mario M. Cuomo, 1992, The Crime Victim in a System of Criminal Justice, 8 Journal of Civil Rights and Economic Development, 5–6.

    New York Office of Victim Service (OVS), 2020, Victim Compensation, Available from: https://ovs.ny.gov/victim-compensation.
33. David L. Roland, 1989, Progress in the Victim Reform Movement: No Longer the "Forgotten Victim", 17 Pepperdine Law Review, 36.
34. Victims of Crime Act: (42 U.S.C. § 10601, 1984).
35. Office for Victims of Crime, 2020, About OVC, Available from: https://ovc.ojp.gov/about-ovc.

Office for Victims of Crime, 2020, Crime Victims Fund, Available from: https://ovc.ojp.gov/sites/g/files/xyckuh226/files/pubs/crimevictimsfundfs/index.html.

36. Melissa L. Jarrell and Joshua Ozymy, 2012, Real crime, real victims: environmental crime victims and the Crime Victims' Rights Act (CVRA), 58 Crime, Law and Social Change, 375.
37. United States v. BP Products North America Inc. (610 F. Supp. 2d 655 2009).

    United States v. CITGO Petroleum Corp 801 F.3d 477 (5th Cir. 2015).
38. For a broader discussion of restorative justice approaches to environmental crime victims, see:

    Rob White, 2014, Eco-Justice and Problem-Solving Approaches to Environmental Crime and Victimisation, In Spapens, et al. (Eds). Environmental Crime and Its Victims, UK: Routledge.

    Matthew Hall, 2014, Victims of Environmental Crime: Routes for Recognition, Restitution and Redress, In Spapens, et al. (Eds.) Environmental Crime and Its Victims, UK: Routledge.

Index

A
Ashland Oil, 51, 64, 115, 123
Avila, Mark, 55, 56, 62, 66

B
Bazan, Arnoldo Carrillo, 96, 99–101, 103, 104, 113, 114
Black Elk Energy, 53, 65

C
CAA, 3, 11
Callihan, Charles Ferris, 4, 11, 81, 82, 87, 113
Cassell, Paul G., 139
CERCLA, 2, 49, 50, 71, 78, 109, 117
CWA, 3, 11, 23, 24, 46–48, 52, 53, 57–59, 71, 72, 76, 77, 97, 108, 109, 111, 120, 121, 130

D
Devaney, Earl E., 10, 28, 40
DOJ, 3, 4, 8, 16, 24, 25, 28, 39, 40, 44, 61, 63, 132, 134, 138
DOJ-ECS, 16, 25, 40, 44, 132, 138

E
Elias, Alan, 56, 57, 60, 66, 113, 114, 122
environmental victimization, 19, 21, 126, 127, 129
EPA, 2–4, 8–11, 16, 23–25, 27, 28, 35–41, 44–47, 50, 55, 57–63, 66, 67, 70–74, 80–86, 90–95, 99–102, 106, 108, 110, 112–116, 118–120, 127–132, 134–138
Equilon Pipeline, 77, 80–82, 86, 87, 113

J. Ozymy and M. Jarrell Ozymy, *The U.S. Administrative State and the Protection of Environmental Crime Victims*, Palgrave Studies in Green Criminology, https://doi.org/10.1007/978-3-031-62680-7

F
FIFRA, 23, 36, 49, 50, 71, 73, 75, 76, 81, 89, 91, 94, 95, 108, 109, 118, 119, 130, 135

H
Halliburton Energy Services, 53, 65
Honeywell International, 48, 58, 59, 62, 67, 113, 114

K
Kimber, Martin, 95, 99, 100, 103, 104, 113

L
Lacey Act, 24
Lazarus, Richard J., 10, 27, 38–40, 136
Lynch, Michael J., 30–35, 40, 138

M
McMurry, Robert I., 10, 11, 28, 39–41
Mintz, Joel A., 40, 41, 135
Monsanto, 118, 119, 123
Motiva Corporation, 59, 67
Mushal, Raymond W., 27, 39, 138

N
NEPA, 23
Norfolk Southern, 2–4, 9, 10

O
Orkin Exterminating Company, 75, 80, 85, 87, 113
OSHA, 2, 9, 44, 55, 57

P
The Pollution Prosecution Act, 24, 39, 131, 136
POTW, 62, 80, 97, 114
Public Employees for Environmental Responsibility, 28, 39, 136, 137

R
RCRA, 3, 11, 23–25, 28, 38, 46–50, 52, 55, 56, 59, 71, 75, 76, 91, 93, 108, 109, 111, 119, 121, 130
Rivers and Harbors Act, 24

S
Southern Union Company, 75, 82, 85, 87, 113
Starr, Judson W., 136

T
Terminix, 75, 85, 86
TSCA, 23, 71, 74, 91, 94–97, 108, 109
Turner, David Eugene, 97, 99–101, 104, 113

U
Uhlmann, David M., 10, 30, 41, 136

The manufacturer's authorised representative in the EU is Springer Nature Customer Service Centre GmbH, Europaplatz 3, 69115 Heidelberg, Germany. If you have any concerns regarding our products, please contact ProductSafety@springernature.com

Printed and bound by CPI Group (UK) Ltd, Croydon, CR0 4YY
01/10/2026
02225978-0011